MATHEMATICS
AND ITS
CREATORS

EDMUND ISAKOV, PH.D.

Mathematics and ItsCreators
Copyright © 2022 by Edmund Isakov, Ph.D.. All rights reserved.

No part of this book may be used or reproduced in any manner whatsoever without written permission, except in the case of brief quotations embodied in critical articles and reviews. For more information, e-mail all inquiries to info@mindstirmedia.com.

Published by Mindstir Media, LLC
45 Lafayette Rd | Suite 181| North Hampton, NH 03862 | USA
1.800.767.0531 | www.mindstirmedia.com

Printed in the United States of America
ISBN-13: 978-1-958729-94-6

This book is dedicated to high school and college students as well as all those who love mathematics.

The main purpose of man on earth is to receive and give knowledge. Knowledge is the only wealth that is multiplied when we give it away.

Boris Roland

If people had not created mathematics, we may never have progressed beyond the Stone Age use of fire.

Edmund Isakov

CONTENTS

Preface ix
Introduction xi

Chapter 1: Prehistoric Mathematics 1
Chapter 2: Babylonian Mathematics 3
Chapter 3: Ancient Egyptian Mathematics 5
3.1 The Moscow Mathematical Papyrus 6
3.2 The Berlin Papyrus 6619 8
3.3 The Rhind Mathematical Papyrus 10

Chapter 4: Greek Mathematics 25
4.1 Thales of Miletus (c. 624–c. 546 BC) 26
4.2 Pythagoras (c. 570–c. 495 BC) 28
4.3 Plato (c. 428–c. 348 BC) 32
4.4 Eudoxus (408–355 BC) 34
4.5 Euclid (c. 330–c. 260 BC) 35
4.6 Archimedes (c. 287–212 BC) 39
4.7 Apollonius (c. 262–c. 190 BC) 45
4.8 Hipparchus (c. 190–c. 120 BC) 47
4.9 Heron (c. 10–75 c. AD) 51
4.10 Ptolemy (c. 85–c. 165 AD) 55
4.11 Diophantus (c. 200–c. 284 AD) 57
4.12 Hypatia of Alexandria (c. 370–415 AD) 59

Chapter 5: Chinese Mathematics 63
5.1 Liu Xin (c. 50 BC–23 AD) 65
5.2 Zhang Heng (78–139 AD) 66
5.3 Liu Hui (c. 225–c. 295 AD) 71

5.4 Zu Chongzhi (429–500 AD) 77
5.5 Yang Hui (1238–1298) 80

Chapter 6: Indian mathematics 83
6.1 Aryabhata (476–550 AD) 84
6.2 Varahamihira (505–587 AD) 86
6.3 Brahmagupta (597–668 AD) 87
6.4 Bhāskara I (c. 600–c. 680 AD) 91
6.5 Mahavira (c. 800–c. 870 AD) 95
6.6 Bhaskara II (1114–1185) 99
6.7 Madhava of Sangamagrama (1340–1425) 103
6.8 Nilakantha Somayaji (1444–1544) 108
6.9 Jyesthadeva (1500–1575) 110

Chapter 7: Islamic mathematics 113
7.1 Muhammad ibn Musa al-Khwarizmi (c. 780–c. 850) 114
7.2 Abu Kamil Shuja ibn Aslam (c. 850–c. 930) 117
7.3 Al-Karaji (953–1029) 119
7.4 Ibn al-Haytham (c. 965–c. 1040) 123
7.5 Omar Khayyam (May 18, 1048–December 4, 1131) 128
7.6 Al-Samawal al-Maghribi (c. 1130–c. 1180) 131
7.7 Nasir al-Din Tusi (February 17, 1201–June 25, 1274) 133
7.8 Jamshid al-Kashi (1380–June 22, 1429) 135

Chapter 8: Medieval European Mathematics 139
8.1 Leonardo Fibonacci (c. 1175–c. 1250) 140
8.2 Thomas Bradwardine (c. 1290–1349) 145
8.3 William Heytesbury (1313–1372/1373) 146
8.4 Nicole Oresme (c. 1320/5–1382) 148

Chapter 9: Renaissance mathematics 151
9.1 Piero Francesca (c. 1415–1492) 152
9.2 Luca Pacioli (c. 1447–1517) 153
9.3 Scipione Del Ferro (1465–1526) 155
9.4 Niccolò Fontana Tartaglia (1500–1557) 158
9.5 Gerolamo Cardano (1501–1576) 160
9.6 Rafael Bombelli (1526–1572) 163

9.7 François Viète (1540–1603) 164
9.8 Bartholomeo Pitiscus (1561–1613) 167

**Chapter 10: Mathematics During the Scientific
           Revolution (17th and 19th centuries) 169**
10.1 John Napier (1550–1617) 170
10.2 René Descartes (1596–1650) 174
10.3 Pierre de Fermat (1601–1665) 177
10.4 Blaise Pascal (1623–1662) 181
10.5 Gottfried Wilhelm Leibniz (1646–1716) 184
10.6 Isaac Newton (1643–1727) 188
10.7 Leonhard Euler (1707–1783) 191
10.8 Joseph-Louis Lagrange (1736–1813) 197
10.9 Pierre-Simon Laplace (1749–1827) 202

Chapter 11: Modern Mathematics (19th –21st centuries) 205
11.1 Carl Friedrich Gauss (1777–1855) 207
11.2 Augustin-Louis Cauchy (1789–1857) 215
11.3 Nikolai Ivanovich Lobachevsky (1792–1856) 218
11.4 Niels Henrik Abel (1802–1829) 222
11.5 Évariste Galois (1811–1832) 226
11.6 János Bolyai (1802–1860) 228
11.7 Bernhard Riemann (1826–1866) 231
11.8 Sofia Vasilyevna Kovalevskaya (1850–1891) 234
11.9 Henri Poincaré (1854–1912) 237
11.10 Georg Cantor (1845–1918) 242
11.11 David Hilbert (1862–1943) 246
11.12 Srinivasa Ramanujan (1887–1920) 248
11.13 Andrey Nikolaevich Kolmogorov (1903–1987) 255
11.14 Grigori Yakovlevich Perelman (born on June 13, 1966) 260
11.15 John Forbes Nash, Jr. (1928–2015) 264

Appendix 1: International Students Performance on
             Mathematics in 2012 269
Appendix 2: International Students Performance on
             Mathematics in 2015 271
Appendix 3: Babylonian numerals 273

Appendix 4: Vladimir Semyonovich Golenishchev (1856–1947) 275
Appendix 5: Vasily Vasilievich Struve (1889–1965) 277
Appendix 6: Alexander Henry Rhind (1833–1863) 279
Appendix 7: James Garfield and the Pythagorean Theorem 281
Appendix 8: Fibonacci Numbers in Pythagorean Triples 285
Appendix 9: Sir Isaac Newton 289

References 297
Afterword 303
About the Author 305

PREFACE

The author wrote this book because he loves mathematics. He hopes that young people and adults, who had read this book, will appreciate mathematics as one of the necessary sciences.

This book is about mathematicians from Greece, China, India, the Islamic Empire, Italy, France, Germany, Scotland, England, Switzerland, Russia, Norway, Hungary, and USA.

By reading this book, you will learn that Isaac Newton (1643–1727), one of the most influential scientists of all time, was a mediocre student. Helped the "happy" case. One of his peers beat him. Direct revenge was excluded since the enemy was stronger than the defeated. Isaac decided to surpass the opponent in training and began to study more diligently and became the best student at the school. Somebody said about the boy who beat Newton: "Nobody acted better with his fists!"

Another genius, Carl Friedrich Gauss (1777–1855), had an exceptional influence in many fields of mathematics and is ranked among of the most influential mathematicians. He was a child prodigy. At 3 years old, he was able to read, write, and even corrected his father's arithmetic. When Carl was in the third grade, the mathematics teacher gave his students the task: to calculate the sum of numbers from 1 to 100. He believed that solution to this problem would keep students busy until the end of the lesson. The teacher was amazed when Carl solved this problem within seconds and brought the correct result. Carl noticed that the pairwise sums from the opposite ends are the same:

$1 + 100 = 101$, $2 + 99 = 101$, etc., and instantly got the result: $50 \times 101 = 5050$.

Andrey Nikolaevich Kolmogorov (1903–1987) was a Soviet mathematician who made significant contribution to a variety of mathematical fields, classical mechanics, and algorithmic information theory. He was one of the greatest mathematicians of the 20th century. At the age of five, Andrey noticed the regularity in the sum of the series of odd numbers:

$$1 = 1 = 1^2$$
$$1 + 3 = 4 = 2^2$$
$$1 + 3 + 5 = 9 = 3^2$$
$$1 + 3 + 5 + 7 = 16 = 4^2$$
$$1 + 3 + 5 + 7 + 9 = 25 = 5^2$$

INTRODUCTION

The Program for International Student Assessment (PISA) is a worldwide study by the Organization for Economic Co-operation and Development (OECD) in member and non-member nations of 15-year-old school students' scholastic performance on mathematics, science, and reading. It was first performed in 2000 and then repeated every three years. It is done for improving education policies and results. It measures problem solving and knowledge in daily life (Ref 1).

In 2012, over half a million students, representing 34 OECD countries and 31 non-member nations, took the internationally agreed two-hour test. Students were assessed in science, mathematics, reading, collaborative problem solving, and financial literacy. The United States delegated 6,111 students from 161 schools. The results in mathematics were published in December 2014 (Appendix 1).

In 2015, students from 70 countries took the same test. The USA delegated 5,712 students from 177 schools.

Average scores of 15-year-old students on mathematics literacy scale from the top ranking, mid-ranking, and low-ranking countries are shown in this table:

Competitions in 2012			Competitions in 2015		
Rank	Country	Average Performance	Rank	Country	Average Performance
1	Shanghai, China	613	1	Singapore	564

XI

2	Singapore	573	2	Hong Kong, China	548
3	Hong Kong, China	561	3	Macau, China	544
4	Taiwan	560	4	Chinese Taipei	542
5	South Korea	554	5	Japan	532
6	Macau, China	538	6	B-S-J-G, China*⁾	531
34	Russia	482	39	Israel	470
36	United States	481	39	United States	470
37	Lithuania	479	41	Croatia	464
62	Qatar	376	62	Peru	387
64	Indonesia	375	63	Indonesia	386
65	Peru	368	70	Dominican Republic	328

*⁾ China: B – Beijing, S – Shanghai, J – Jiangsu, and G – Guangdong

The results of competitions among all countries are shown in Appendix 1 (p. 228) and Appendix 2 (p. 229).

Performance of the students from the United States was very disappointing. Let me express my opinion. Normally, 15-year-old students in the USA public schools are in Grade 9. They take mathematics "Algebra 1" as a required curriculum in high school. "Algebra 1" contains 37 sections: *Numbers, Ratios and Proportions, Percent, Measurements, Coordinate graphs, Solve equations*, and more. Surprisingly, *Geometry* and *Trigonometry* are also included in "Algebra 1." I think this is very bad: geometry and trigonometry should be studied separately from algebra.

In the Soviet Union study of algebra and geometry began from 6th grade. The textbooks were: *Algebra for 6th Grade of middle schools* and *Geometry for 6th Grade of middle schools*. These books, written by the Russian/Soviet mathematician Andrei Petrovich Kiselëv (1852–1940), were official textbooks endorsed by the Ministry of Education for use in all schools of the Soviet Union.

During school years, my favorite subjects were mathematics, physics, chemistry, and astronomy. My teachers inspired me to be a good student and strive to have a good knowledge in those branches of science. My desire to study sciences and gain more knowledge has never stopped.

In the age of new technologies, Google and Wikipedia, everyone can find out what is happening around us, across the world, on planets, the solar system, the universe, etc.

National Science Foundation conducted the survey and over 2,200 people in the United States were asked few questions from basic science. Only **74%** of the participants knew that **Earth is orbiting around the Sun**. It means that **one** in **four** Americans still believe that **the Sun is orbiting around our Earth**. According to Gallup poll, **4** in **10** Americans believe Earth was created by God 10,000 years ago.

Unfortunately, all available means for learning and expanding our knowledge are not fully employed. Maybe, that is why we have such a gloomy statistic in the USA:

- Over 1.2 million students drop out of high school every year
- About 25% of high school freshmen fail to graduate from high school on time
- Nearly two thousand high schools graduate only less than 60% of their students
- High school dropouts commit about 75% of crime.

This book contains the following periods of mathematics: Prehistoric, Babylonian, Egyptian, Greek, Chinese, Indian, Islamic, Medieval European, Renaissance, Mathematics during the Scientific Revolution (17th and 18th centuries), and Modern Mathematics (19th, 20th, and 21st centuries).

The most difficult task for the author was the choice of mathematicians and how many of them. He included in the book 70 mathematicians who made the greatest contribution to mathematics.

"In order to understand the universe, you must know the language in which it is written. And that language is mathematics," said

Galileo Galilei (1564–1642). He was an Italian astronomer, physicist, and engineer.

"Mathematics is the most beautiful and most powerful creation of the human spirit," said Stefan Banach (1892–1945). He was a Polish mathematician.

CHAPTER 1

PREHISTORIC MATHEMATICS

The origin of mathematics took place about 20,000 years ago. It was based on the concepts of number, magnitude, and form. It was a necessity of everyday life in hunter-gatherer tribes. The idea of the "number" gave them ability to count. This idea is supported by the existence of languages, which preserve the distinction between "one," "two," and "many," but not of numbers larger than two. Such "mathematics" was a prefiguration of arithmetic and geometry.

Egyptians of the 5th millennium BC pictorially represented geometric designs.

Gigantic monuments in England and Scotland, dating from the 3rd millennium BC, incorporate geometric ideas such as circles, ellipses, and Pythagorean triples in their design.

All of the above are disputed, and the currently oldest undisputed mathematical documents are from Babylonian and dynastic Egyptian sources (Ref 2).

CHAPTER 2

BABYLONIAN MATHEMATICS

Babylonian mathematics means any mathematics of the peoples of Mesopotamia (modern Southern Iraq). The earlier evidence of written mathematics dates back to the ancient Sumerians, who built the earliest civilization in Mesopotamia. About 5000 years ago, the Sumerians developed numerals, multiplication tables on clay tablets, dealt with geometrical exercises, and division problems. Babylonian mathematics were written using sexagesimal (base-60) numerals[1]). Sexagesimal system is still in use: 60 second in 1 minute, 60 minutes in 1 hour, and 360 degrees in a circle. Minutes and seconds of arc are used to denote fractions of a degree. Probably, the sexagesimal system was chosen because 60 can be evenly divided by 2, 3, 4, 5, 6, 10, 12, 15, 20, and 30. Unlike the Egyptians, Greeks, and Romans, the Babylonians had a true place-value system, where digits written in the left column represented larger values, much as in the decimal system. Babylonian mathematics also covered fractions, algebra, quadratic and cubic equations. It was a remarkable <u>achievement for the time (Ref 2).</u>

---

[1] The Babylonian numerals (Ref 3) were written in cuneiform using a wedge-tipped reed stylus to make a mark on a soft clay tablet, which was exposed in sunlight to create a permanent record. Detailed information on the Babylonian numerals is provided in Appendix 3.

CHAPTER 3

ANCIENT EGYPTIAN MATHEMATICS

Ancient Egyptian mathematics (Ref 2) is the mathematics that was developed and used in Ancient Egypt (c. 3000 to c. 300 BC). The ancient Egyptians utilized a numeral system (numeration based on multiples of ten, rounded off to the higher power, written in hieroglyphs) for counting and solving written mathematical problems, involving multiplication and fractions.

Written evidence of the use of mathematics dates back to at least 3000 BC with the ivory labels found in Tomb U-j at Abydos[1]. These labels appear to have been used as tags for grave goods and some are inscribed with numbers. Further evidence of the use of the base 10 number system can be found on the Narmer Macehead[2] which depicts offerings of 400,000 oxen, 1,422,000 goats and 120,000 prisoners.

The evidence of the use of mathematics in the Old Kingdom[3] is scarce, but can be deduced from inscriptions on a wall near a mastaba[4] in Meidum[5] which gives guidelines for the slope of the mastaba. The

---

[1] Abydos is one of the oldest cities of ancient Egypt
[2] The Narmer Macehead is an ancient Egyptian decorative.
[3] The Old Kingdom is the period in the third millennium (2686–2181 BC) also known as the "Age of the Pyramids."
[4] A mastaba is a type of ancient Egyptian tomb in the form of a flat-roofed, rectangular structure with inward sloping sides.
[5] Meidum, is an archaeological site in Lower Egypt. It contains a large pyramid and several mud-brick mastabas.

lines in the diagram are spaced at a <u>distance of one cubit and show the use of that unit of measurement.</u>

Evidence for Egyptian mathematics is limited to a scarce amount of surviving sources written on papyri. From these texts it is known that ancient Egyptians understood concepts of geometry, such as determining the surface area and volume of three-dimensional shapes useful for architectural engineering.

Significant mathematical documents are the papyruses from the Middle Kingdom period of ancient Egypt dated to 2000–1700 BC.

There are three major mathematical documents: the Moscow Mathematical Papyrus (1890 BC), the Berlin Papyrus (1800 BC), and the Rhind Mathematical Papyrus, dated to c.1650 BC.

3.1 The Moscow Mathematical Papyrus

The Moscow Mathematical Papyrus, known as the Golenishchev Mathematical Papyrus (Ref 4), named after its first owner, Egyptologist Vladimir Semyonovich Golenishchev (his biography is in Appendix 4). Currently, this papyrus is exhibited in the Pushkin State Museum of Fine Arts in Moscow, Russian Federation.

It is a well-known mathematical papyrus, approximately 5.5 m (18.1 ft) long and varying between 3.8 and 7.6 cm (1.5 and 3.0 in.) wide. Its format was divided into 25 problems. Translations and solutions of the problems were made by the Soviet Orientalist Vasily Vasilievich Struve in 1930 (for his biography see Appendix 5).

The problems in the Moscow Papyrus follow no particular order. The papyrus is well known for finding unknown quantities, geometry problems, and problems that are more common in nature.

Aha problems

Aha problems involve finding unknown quantities (referred to as Aha) if the sum of the quantity and part(s) of it are given. For instance, problem 19 asks one to calculate a quantity taken 1 and 1/2 times and added to 4 make 10. In modern mathematical notation, this is an equation with one unknown:

$$3/2\, x + 4 = 10$$

By simplifying the left-hand part of the above equation, we obtain:

$$3/2\, x = (10 - 4)$$
$$3/2\, x = 6$$
$$x = 6/(3/2) = 6 \times 2/3 = 4$$

Geometry problems

Seven of the twenty-five problems and range from computing areas of triangles, to finding the surface area of a hemisphere (problem 10) and finding the volume of a frustum of square pyramid (problem 14). The sketch of frustum of square pyramid is shown below.

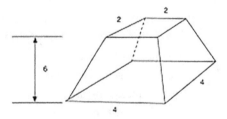

The text of problem 14: "If you are told: a truncated pyramid of 6 for the vertical height by 4 on the base, by 2 on the top. You are to square the 4 - result 16. You are to double 4 - result 8. You are to square this 2 - result 4. You are to add the 16 and the 8 and the 4 - result 28. You are to take 1/3 of 6 - result 2. You are to take 28 twice - result 56. You will find right."

In modern mathematical notation, the volume V is:

$$V = \tfrac{1}{3} h(a^2 + ab + b^2)$$

where

h = 6 (vertical height), a = 2 (length of a top side), b = 4 (length of a bottom side):

$$V = 1/3 \times 6 \, (2^2 + 2 \times 4 + 4^2) = 2 \, (4 + 8 + 16) = 2 \times 28 = 56.$$

The solution to the problem indicates that the Egyptians knew the correct formula for obtaining the volume of a truncated pyramid.

Researchers have speculated how the Egyptians might have applied the formula for the volume of a frustum of a square pyramid if the derivation of this formula is not given in the papyrus.

3.2 The Berlin Papyrus 6619

The Berlin Papyrus 6619 (Ref 5), commonly known as the Berlin Papyrus, is one of the primary sources of ancient Egyptian mathematics. The Berlin papyrus contains two mathematical problems that were translated and published by a German Egyptologist Hans Schack-Schackenburg (1853–1905) in 1900 and 1902.

Problem 1. "You are told the area of a square of 100 square cubits is equal to that of two smaller squares, the side of one square is 1/2 + 1/4 of the other. What are the sides of the two unknown squares?"

In modern mathematical notation, this problem is defined by two equations in two unknowns:

$$x^2 + y^2 = 100 \qquad (1)$$
$$x = (1/2 + 1/4) \, y = (3/4) \, y \qquad (2)$$

Substitution of x in equation (1) by x in equation (2) gives:

$$[(3/4) \, y]^2 + y^2 = 100$$
$$(9/16) \, y^2 + y^2 = 100$$
$$y^2 \, (9/16 + 1) = 100$$
$$y^2 \, (9/16 + 16/16) = 100$$
$$y^2 \, (25/16) = 100$$

$$y^2 = 100 / (25 / 16)$$
$$y^2 = (100 \times 16) / 25 = 1600/25$$
$$y^2 = 64$$
$$y = \sqrt{64} = 8$$
$$x = (3/4) \times y = (3/4) \times 8$$
$$x = 24/4 = 6$$
Result: $x^2 + y^2 = 6^2 + 8^2 = 36 + 64 = 100$

Problem 2. "You are told the area of a square of 400 square cubits is equal to that of two smaller squares, the side of one square is 1/2 + 1/4 of the other. What are the sides of the two unknown squares?"

In modern mathematical notation, this problem is defined by two equations in two unknowns:

$$x^2 + y2 = 400 \qquad (1)$$
$$x = (1/2 + 1/4)\, y = (3/4)\, y \qquad (2)$$

Solution to this problem is similar to Problem 1. Substitution of x in equation (1) by x in equation (2) gives:

$$((3/4)\, y)^2 + y^2 = 400$$
$$(9/16)\, y^2 + y^2 = 400$$
$$y^2 (9/16 + 1) = 400$$
$$y^2 (9/16 + 16/16) = 400$$
$$y^2 (25/16) = 400$$
$$y^2 = 400 / (25/16)$$
$$y^2 = (400 \times 16) / 25 = 6400 / 25$$
$$y^2 = 256$$
$$y = \sqrt{256} = 16$$
$$x = (3/4) \times y = (3/4) \times 16$$
$$x = 48 / 4 = 12$$
Result: $x^2 + y^2 = 12^2 + 16^2 = 144 + 256 = 400$

3.3 The Rhind Mathematical Papyrus

The Rhind Mathematical Papyrus (Ref 6) is the best example of Egyptian mathematics. Papyrus is named after Alexander Henry Rhind, a Scottish antiquarian (his biography is provided in Appendix 6). Rhind purchased this papyrus in 1858 in Luxor, Egypt. The papyrus was found during illegal excavation near Ramesseum (the memorial temple of Pharaoh Ramesses). The papyrus dates to around 1650 BC.

In 1865, the British Museum acquired the Rhind Mathematical Papyrus and the Egyptian Mathematical Leather Roll, which also was owned by Alexander H. Rhind. The papyrus began to be transliterated and mathematically translated in the late 19th century.

In 1923, Thomas Eric Peet (1882–1934), English Egyptologist, published the Rhind Mathematical Papyrus, which contains Book I, Book II, and Book III.

Book I

The first part of the Rhind papyrus consists of reference tables and a collection of 20 arithmetic and 20 algebraic problems. The problems start out with simple fractional expressions, followed by completion problems and more involved linear equations (*aha* problems). *Aha* problems involve finding unknown quantities.

The first part of the papyrus is taken up by the $2/n$ table. The fractions $2/n$ for odd n ranging from 3 to 101 are expressed as sums of unit fractions.

For example:

$$2/15 = 1/10 + 1/30$$

The decomposition of $2/n$ into unit fractions is never more than 4 terms long as in this example: $2/101 = 1/101 + 1/202 + 1/303 + 1/606$

The list of fraction expressions contains the numbers 1 through 9 divided by 10. For instance, the division of 7 by 10 is recorded as:

7 divided by 10 yields 2/3 + 1/30

In modern mathematical notation, it is:

$7/10 = 2/3 + 1/30$

Solution:

$2/3 + 1/30 = (20 + 1)/30 = 21/30 = 7/10$

Conclusion: the equation is correct.

Problems 1–6 are compute divisions of a certain number of loaves of bread by 10 men and record the outcome in unit fractions. Problems 7–20 show how to multiply the expressions $1 + 1/2 + 1/4$ and $1 + 2/3 + 1/3$ by different fractions.

Problems 21–23 are problems in completion, which in modern notation is simply a subtraction problem. The problem is solved by the scribe to multiply the entire problem by a least common multiple of the denominators, solving the problem and then turning the values back into fractions.

Problems 24–34 are *aha* problems (involve finding unknown quantities). These problems are linear equations. Problem 32 corresponds (in modern notation) to solving for x:

$$x + 1/3\, x + 1/4\, x = 2 \qquad (1)$$

Simplification of the left-hand part of the equation (1) gives:

$$x + 1/3\, x + 1/4\, x = x\,(1 + 1/3 + 1/4)$$

The least common denominator for equation (1) is: $3 \times 4 = 12$

Adding the fractions in equation (1), gives:

$$x(12/12 + 4/12 + 3/12) = x(19/12) \qquad (2)$$

Rewriting the right-hand part of equation (2):

$$(19/12)x = 2$$

Solution to x in equation (2):

$$x = 2/(19/12) = 2 \times 12/19 = 24/19$$
$$x = 24/19$$

Rewriting the left-hand part of equation (1), gives:

$$24/19 + (1/3 \times 24/19) + (1/4 \times 24/19) =$$
$$24/19 + 24/57 + 24/76 \qquad (3)$$

The least common denominator for fractions of equation (3) is:

$$19 \times 4 \times 3 = 228$$

Multiplying and adding the fractions in the left-hand part of equation (1), gives:

$$[(24 \times 12) + (24 \times 4) + (24 \times 3)]/228 =$$
$$(288 + 96 + 72)/228 =$$
$$456/228 = 2$$

Equation (1) is correct, because its left-hand part and right-hand part are equal.

Problems 35–38 involve divisions of the *hekat* or *heqat*. Hekat was an ancient Egypt volume unit to measure grain, bread, and beer. 1

hekat equals 4.8 liters in today's measurement by the metric unit. It is equal to 1.268 gallons.

Book II

The second part of the Rhind papyrus consists of geometry problems.

Volumes

Problems 41–46 show how to find the volume of both cylindrical and rectangular based granaries. In problem 41 the scribe computes the volume of a cylindrical granary. Diameter (d), height (h), and volume (V) are given in modern mathematical notation by the formula:

$$V = [(1 - 1/9) d]^2 h = (8/9)^2 d^2 h$$

Using $d = 2r$ (r is the radius of a cylinder) and to substitute d by 2r, obtain:

$$V = (8/9)^2 (2r)^2 h = (8/9)^2 2^2 r^2 h$$
$$V = (64/81) \times 4r^2 h = (256/81) r^2 h$$

The improper fraction $256/81 \approx 3.1605$, which approximates $\pi \approx 3.1416$.

About 3700 years ago the ancient Egyptians used the prototype of π-value, which was greater than π by:

$$3.1605 - 3.1416 = 0.0189.$$

It is only 0.60% difference. It is amazing!

Areas

Problems 48–55 show how to compute an assortment of areas. Problem 48 is often commented on as it computes the area of a circle. The scribe compares the area of a circle (approximated by an octagon) and its circumscribing square. Each side is trisected, and the corner triangles are then removed. The resulting octagonal figure approximates the circle. The area of the octagonal figure is:

$$9^2 - 4 \times (1/2 \times 3 \times 3) = 9^2 - 4 \times 9/2 =$$
$$81 - 36/2 = 63 \qquad (1)$$

Next we approximate 63 to 64 and note that $64 = 8^2$, and we get the approximation

$$\pi \times (9/2)^2 \approx 8^2 \qquad (2)$$

Solving for π, we get the approximation $\pi \approx 256/81 \approx 3.1605$ (the approximation has an error of 0.0189). "The octagonal figure, whose area is easily calculated, accurately approximates the area of the circle is just plain good luck."

The equations (1) and (2) are written in the modern mathematical notation. It is not clear why the given numbers are used in these equations. Unfortunately, there is no dimension of the octagon side.

There are two possibilities: an octagon inscribed in a circle (Case 1), and a circle is inscribed in an octagon (Case 2).

Case 1: An octagon inscribed in a circle

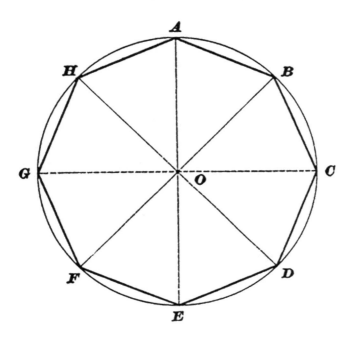

Octagon is divided into 8 isosceles triangles that have one vertex at the center of the circle and the other vertices on the edge of the circle. The measure of each angle at the center of the circle is 45 degrees (360 degrees divided by 8).

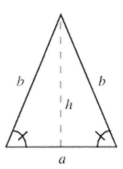

Area of each triangle is: $A_T = 1/2\ a \times h$, where a is a base and h is a height. Angles in this triangle are vertex angle is 45° and base angles are 67.5° each. Assuming that a base of a triangle (a) equals 1 unit, then a height is:

$$h = 1/2\ a \times \tan 67.5° = 1/2\ a \times 2.4142 = 1.2071a$$

Area of the triangle is:

$$A_T = 1/2\ a \times h = 1/2\ a \times 1.2071a = 0.60355a^2$$

Area of the octagon inscribed in the circle (A_O) equals the sum of the areas of 8 isosceles triangles:

$$A_O = A_T \times 8 = 0.60355a^2 \times 8 = 4.828a^2.$$

Area of the circumscribed circle (A_C) is:

$$A_C = \pi \times b^2$$

Where b is a circumradius (8 circumradii are: from 0A to 0H shown in the above octagon) and the sides of a triangle (b are shown in the above isosceles triangle):

$$b^2 = (a/2)^2 + h^2 = (a/2)^2 + (1.2071a)^2 =$$
$$0.25a^2 + 1.4571a^2 = 1.7071a^2$$

$$A_C = \pi \times b^2 = \pi \times 1.7071a^2 = 5.363a^2$$

Area of the circumscribed circle is greater than that of the inscribed octagon by:

$$A_C - A_O = 5.363a^2 - 4.828a^2 = 0.535a^2,\ \text{or}$$
$$0.535a^2 \times 100/4.828a^2 \approx 11\%$$

Case 2: A circle inscribed in an octagon

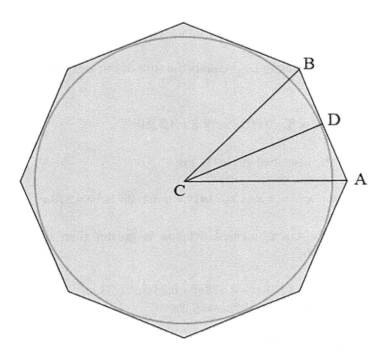

Area of an isosceles triangle ACB (A_T) is calculated similar to that in Case 1.

$$A_T = 1/2 \times AB \times CD = 1/2 \times a \times h$$

AB is a base of a triangle and a side of an octagon (a). CD is a height of a triangle and a radius of an inscribed circle (h).

$$A_T = 1/2 \times a \times h$$

Assuming that $a = 1$ unit, then a height is:

$$h = AD \times \tan \angle CAD = 1/2\, a \times \tan 67.5° = 1/2\, a \times 2.4142 = 1.2071a$$

Area of an isosceles triangle is:

$$A_T = 1/2 \times a \times 1.2071a = 0.60355a^2$$

Area of the octagon (A_O) equals the sum of the areas of 8 isosceles triangles:

$$A_O = A_T \times 8 = 0.60355a^2 \times 8 = 4.828a^2$$

Area of the inscribed circle (A_C) is:

$$A_C = \pi \times h^2 = \pi \times (1.2071a)^2 = \pi \times 1.4571a^2 = 4.578a^2$$

Area of the circumscribed octagon is greater than that of the inscribed circle by:

$$A_O - A_C = 4.828a^2 - 4.578a^2 = 0.25a^2,$$
or $0.25a^2 \times 100/4.828a^2 \approx 5.2\%$

Calculations in the above cases showed that the statement in the Rhind Mathematical Papyrus, "the octagonal figure, whose area is easily calculated, accurately approximates the area of the circle" is wrong.

Pyramids

The final five problems are related to the slopes of pyramids. The inclination of the triangular faces of a right pyramid was measured by seked (or seqed). The system was based on the Egyptian's length measure known as the royal cubit. The royal cubit was subdivided into seven palms and each palm was further divided into four digits. The inclination of measured slopes was expressed as the number of palms and digits moved horizontally for each royal cubit rise.

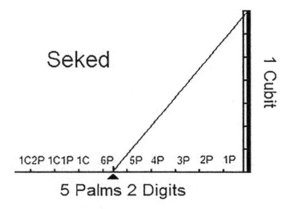

The most famous of all pyramids of Egypt is the Great Pyramid of Giza built around 2,550 B.C.

Based on the surveys of this structure that have been carried out by Finders Petrie[6] and others, the slopes of the faces were a seked of 5½, or 5 palms and 2 digits (see figure above), which equates to a slope of 51.84° from the horizontal, using the modern 360 degrees system.

The final five problems in Book II are related to the slopes of pyramids. One seked problem is:

"If a pyramid is 250 cubits high (h) and the side of its base 360 cubits long (b), what is its seked?"

The solution to the problem is given as the ratio of half the side of the base of the pyramid to its height, or the run-to-rise ratio of its face.

---

[6] Sir William Matthew Flinders Petrie, Fellow of the Royal Society (1853–1942), commonly known as Flinders Petrie, was an English Egyptologist and a pioneer of systematic methodology in archaeology and preservation of artifacts. He held the first chair of Egyptology in the United Kingdom. He excavated many of the most important archaeological sites in Egypt in conjunction with his wife, Hilda Petrie. Flinders Petrie developed the system of dating layers based on pottery and ceramic findings.

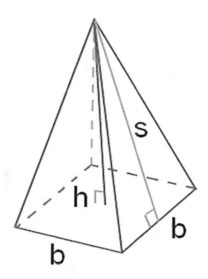

In modern mathematical notation, the quantity for the seked is the cotangent of the angle to the base of the pyramid and its face.

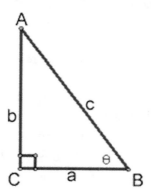

This right-angled triangle is a section of the above right pyramid. The section goes through the top of pyramid (apex) perpendicularly to the base, where: AC is a height of a pyramid (h = 250, AC = 250), AB is a height of a lateral face (s), BC is half the side of base (b = 360, BC =

180), and ÐΘ is the angle to the base of the pyramid and its face. This angle is calculated as shown below:

The value of ÐΘ is:

$$\text{Cotangent } \Theta = BC/AC = 180/250 = 0.72$$
$$\text{ÐΘ} = \text{arccotangent } 0.72 = 54°14¢46^2 = 54.2461°$$

Book III

The third part of the Rhind papyrus consists of the reminder of the 84 problems. Problem 61 consists of 2 parts. Part 1 contains multiplications of fractions. Part 2 gives a general expression for computing $2/3$ of $1/n$, where n is odd number. In modern mathematical notation, the equation is:

$$2/3n = 1/2n + 1/6n$$

To check the accuracy of this equation, let perform calculations for two odd numbers.

<u>Example 1</u>, $n = 5$:

$$\underline{2/15} = 1/10 + 1/30$$

Adding the fractions of the right-hand part of the equation gives:

$$1/10 + 1/30 = 3/30 + 1/30 = 4/30 = \underline{2/15}$$

<u>Example 2</u>, $n = 201$:

$$\underline{2/603} = 1/402 + 1/1206$$

Adding the fractions of the right-hand part of the equation gives:

$$1/402 + 1/1206 = 3/1206 + 1/1206 = 4/1206 = \underline{2/603}$$

Since the left-hand part and the right-hand part are equal, the equation is accurate.

Let's see if the equation is applicable for even numbers.

Example 3, $n = 10$:

$$2/30 = 1/20 + 1/60$$

Adding the fractions of the right-hand part of the equation gives:

$$1/20 + 1/60 = 3/60 + 1/60 = 4/60 = \underline{2/30}$$

Example 4, $n = 200$:

$$\underline{2/600} = 1/400 + 1/1200$$

Adding the fractions of the right-hand part of the equation gives:

$$1/400 + 1/1200 = 3/1200 + 1/1200 = 4/1200 = \underline{2/600}$$

As can be seen, this equation is also applicable for even numbers.

Problems 62–68 are general problems of an algebraic nature. Problems 69–78 are all *pefsu* problems in some form or another (*pefsu* is the number loaves of bread or jugs of beer divided by number of heqats of grain; *pefsu* measures the strength of the beer or bread made from a heqat of grain. A higher *pefsu* number means weaker beer or bread). They involve computations regarding the strength of bread and or beer.

Problem 79 sums five terms in a geometric progression[7]. It is a multiple of 7 riddle, which would have been written in the Medieval[8] era as, "Going to St. Ives"[9] problem.

Problems 80 and 81 compute the Eye[10] of Horus[11] fractions of hekats (heqats).

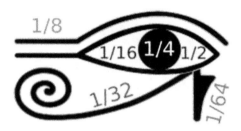

The right side of the eye = 1/2, the pupil = 1/4, the eyebrow = 1/8, the left side of the eye = 1/16, the curved tail = 1/32, and the teardrop = 1/64.

Problem 81 is followed by a table contains fractions as portions of square.

---

[7] Geometric progression is a sequence of numbers where each term after the first is found by multiplying the previous one by a fixed, non-zero number called the common ratio. For example, the sequence 3, 6, 12, 24, 48, 96…is a geometric progression with common ratio 2.

[8] In European history, the medieval period lasted from the 5th to the 15th century. It began with the collapse of the Western Roman Empire and merged into the Renaissance and the Age of Discovery.

[9] "Going to St. Ives" is a traditional English-language nursery rhyme in the form of a riddle.

[10] The Eye of Horus is an ancient Egyptian symbol of protection, royal power and good health. Arithmetic values represented by parts of the Eye of Horus.

[11] Horus is one of the most significant gods and goddesses in ancient Egyptian religion. Horus was usually depicted as a falcon-headed man wearing the double crown or a red and white crown, as a symbol of kingship over the entire kingdom of Egypt.

CHAPTER 4

GREEK MATHEMATICS

The origin of Greek mathematics (Ref 2) is not well documented. The earliest advanced civilization in Greece and in Europe were the Minoan[1] and later Mycenaean[2] civilizations, both of which flourished during the 2nd millennium BC. While these civilizations possessed writing and were capable of advanced engineering, including four-story palaces with drainage and beehive tombs, they left <u>behind no mathematical documents.</u>

Greek mathematics refers to mathematics texts and advances written in Greek, developed from the 7th century BC to the 4th century AD around the shores of the Eastern Mediterranean. Greek mathematicians lived in cities spread over the entire Eastern Mediterranean, from Italy to North Africa, but were united by culture and language.

Greek mathematics of the period following Alexander the Great (356–323 BC) is sometimes called Hellenistic mathematics (period from 323 BC to 31 BC). The word "mathematics" itself derives from the ancient Greek *máthēma*, meaning "subject of instruction." The study of mathematics for its own sake and the use of generalized mathematical theories and proofs is the key difference between Greek mathematics and those of preceding civilizations.

---

[1] Minoan civilization was an Aegean Bronze Age civilization on the island of Crete and other Aegean islands, which flourished from about 2600 to 1600 BC.
[2] Mycenaean civilization was the last phase of the Bronze Age in Ancient Greece, spanning the period from approximately 1600–1100 BC.

Greek mathematics was much more sophisticated than the mathematics that had been developed in Babylon and ancient Egypt.

Greek mathematics had begun with Thales of Miletus and Pythagoras of Samos. It is believed that both were inspired by Egyptian and Babylonian mathematics.

4.1 Thales of Miletus (c. 624–c. 546 BC)

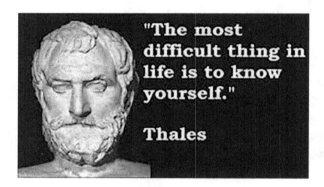

Thales (Ref 7) was a Greek mathematician, philosopher, and astronomer. Aristotle[3] regarded him as the first philosopher in the Greek tradition. Historically he is recognized as the first individual in Western civilization known to have <u>entertained and engaged in scientific thought.</u>

Thales is recognized as a scientist, who broke from understanding the world and universe by mythological explanations for the existence of natural phenomena explained by theories and hypothesis.

Thales used geometry to calculate the height of pyramids and the distance of ships from the shore. He was the first mathematician to use deductive reasoning[4] <u>applied to geometry by deriving four results to Thales' Theorems</u>

---

[3] Aristotle (384–322 BC) was a Greek philosopher and polymath during the Classical period in Ancient Greece.

[4] Deductive reasoning is a logical process in which a conclusion is based on the agreement of multiple premises that are generally assumed to be true. He is the first known individual to whom a mathematical discovery has been attributed.

Reference to Thales was made by Diogenes Laërtius[5], who said that Thales "was the first to inscribe in a circle a right-angle triangle" (Ref 8).

It is believed that Thales learned that an angle inscribed in a semicircle is a right angle during his travels to Babylon. The theorem is named after Thales, because he was said by ancient sources to have been the first to prove the theorem, using his own results that the base angles of an isosceles triangle are equal, and that the sum of angles in a triangle is equal to 180°.

In modern geometrical notation, Thales' Theorem states that if A, B and C are points on a circle where the line AC is a diameter of the circle, then the angle ∠ABC is a right angle.

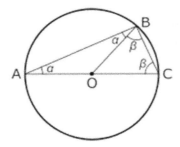

The following facts are used: the sum of the angles in a triangle is equal to 180° and the base angles of an isosceles triangle are equal. Since OA = OB = OC, ΔOAB and ΔOBC are isosceles triangles, and by the equality of the base angles of an isosceles triangle, ∠OBC = ∠OCB and ∠BAO = ∠ABO.

Let's α = ∠BAO and b = ∠OBC. The three internal angles of the ΔABC triangle are α, ($\alpha + \beta$), and β. Since the sum of the angles of a triangle is equal to 180°, we have:

$$\alpha + (\alpha + \beta) + \beta = 180°$$
$$2\alpha + 2\beta = 180°$$

---

[5] Diogenes Laërtius (180–240 AD) was a biographer of the Greek philosophers. Little is known about his life.

$$2(\alpha + \beta) = 180°$$
$$\alpha + \beta = 180°/2 = 90°$$

Q.E.D.

Q.E.D. is an initialism of the Latin phrase *quod erat demonstrandum*, meaning "which is what had to be proven." The phrase is traditionally placed in its abbreviated form at the end of a mathematical proof or philosophical argument.

4.2 Pythagoras (c. 570–c. 495 BC)

Pythagoras (Ref 9) was an Ionian[6] Greek philosopher, mathematician, and has been credited as the founder of the movement called Pythagoreanism[7]. Most of the information about Pythagoras was written down centuries after he lived, so very little reliable information is known about him.

---

[6] The Ionians were one of the four major tribes that the Greeks considered themselves to be divided into during the ancient period.

[7] Pythagoreanism originated in the 6th century BC, based on the teachings and beliefs held by Pythagoras and his followers, the Pythagoreans, who were considerably influenced by mathematics and mysticism.

Pythagoras traveled to Egypt, Greece, and maybe India to learn mathematics, geometry, and astronomy. He established the Pythagorean School whose motto was "All is number," and its doctrine was that mathematics ruled the universe. It was the Pythagoreans who coined the term "mathematics" (from Greek *máthēma*, meaning "knowledge, study, and learning").

Pythagoras is best known for the Pythagorean Theorem, which bears his name. There is some evidence that Babylonian, Mesopotamian, Indian, and Chinese mathematicians all discovered the theorem independently and, in some cases, provided proofs for special cases.

Many of the accomplishments credited to Pythagoras may actually have been accomplishments of his colleagues and successors.

Philosophy associated with Pythagoras was related to mathematics and that numbers were important. It was said that he was the first man to call himself a philosopher, or lover of wisdom, and Pythagorean ideas exercised a marked influence on Plato, and through him, all of Western philosophy.

Theorem states that the square of the hypotenuse c (the side opposite the right The Pythagorean angle) is equal to the sum of the squares of the other two sides. In modern geometrical notation, the theorem can be written as an equation relating the lengths of the sides a, b, and c:

$$a^2 + b^2 = c^2$$

The following proof is (Ref 10).

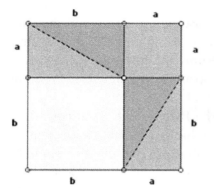 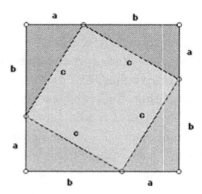

The aria of the left square is:
$(a + b) \times (a + b) = (a + b)^2$ or $4(1/2ab) + a^2 + b^2$

The area of the right square is:
$4(1/2ab) + c^2$

Since the squares have equal areas, we can set them equal:
$4(1/2ab) + a^2 + b^2 = 4(1/2ab) + c^2$

Removing equal quantities from both sides of equation, we have:
$a^2 + b^2 = c^2$

Unique proof of this theorem was done by James Garfield, the President of the United States. Detailed information on his proof is provided in Appendix 7.

The popularity of this theorem became the source of many humorous proverbs. There is a Russian one, "The Pythagorean pants on all sides are equal" (translated by Edmund Isakov from the Russian "Пифагоровы штаны на все стороны равны").

Pythagorean Theorem is ranked № 4 in the list of "The Hundred Greatest Theorems" (Ref 11).

This is a theorem that may have more known proofs than any other. The book *Pythagorean Proposition*, by Elisha Scott Loomis[8]), contains 367 proofs.

---

[8] Elisha Scott Loomis (1852–1940) was an American teacher, mathematician, engineer, writer, and genealogist.

4.3 Plato (c. 428–c. 348 BC)

Plato (Ref 2, 12) was a philosopher and mathematician in Classic Greece (a period in Greece culture lasted from the 5th through 4th centuries BC). Inspired by Pythagoras, Plato founded his Academy in Athens in 387 BC, where he stressed mathematics as a way of understanding more about reality. In particular, he was convinced that geometry was the key to unlocking the secrets of the universe. The sign above the Academy entrance read: "Let no-one ignorant of geometry enter here."

Plato played an important role in encouraging and inspiring Greek intellectuals to study mathematics as well as philosophy. His Academy taught mathematics as a branch of philosophy. The first 10 years of the 15-year course at the Academy, students were involved in study of science and mathematics, including plane and solid geometry, astronomy and harmonics. Plato became known as the "maker of mathematicians," and his Academy boasted some of the most prominent mathematicians of the ancient world.

Plato demanded of his students' accurate definitions, clearly stated assumptions, and logical deductive proof. He insisted that geometric proofs be demonstrated with no aids other than a straight edge and a compass. Plato posed investigation for his students many mathematical problems among which were the Three Classical Problems: squaring the circle, doubling the cube, and trisecting the angle.

As a mathematician, Plato (Ref 12) is best known for his identification of five regular symmetrical 3-dimensional shapes. He called them the basis for the whole universe.

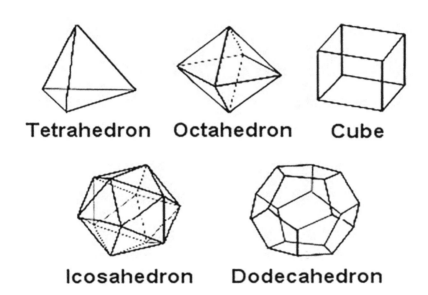

These five shapes have become known as the Platonic Solids: the **tetrahedron** (constructed of 4 regular triangles, which for Plato represented fire), the **octahedron** (composed of 8 regular triangles, represented air), the **cube** (composed of 6 squares, representing earth), the **icosahedron** (composed of 20 regular triangles, represented water), and the **dodecahedron** (made of 12 pentagons, which Plato described as "the god used for arranging the constellations on the whole heaven").

Plato is widely considered the most pivotal figure in the development of philosophy, especially the Western tradition. Unlike nearly all of his philosophical contemporaries, Plato's entire work is believed to have survived intact for over 2,400 years.

Many philosophers around the world agree that Greek philosophy has influenced much of Western culture since its inception.

4.4 Eudoxus (408–355 BC)

Eudoxus (Ref 13) was a mathematician, astronomer, scholar and student of Plato. All of his works are lost, though some fragments are preserved in Hipparchus'[9] commentary on Aratus's[10] poem on astronomy.

Eudoxus developed a method of exhaustion[11], a precursor to the integral calculus. Applying his method, Eudoxus proved such mathematical statements as: areas of circles are to one another as the squares

---

[9] Hipparchus (c.190–c.120 BC) was a Greek astronomer, mathematician, and geographer. He is considered the founder of trigonometry, but the most famous for his incidental discovery of precession of the equinoxes. For more details on him, see Section 4.8.

[10] Aratus (315/310–240 BC) was a Greek poet. His major extant work is a poem *Appearances*, the first half of which is a verse setting of a lost work of the same name by Eudoxus.

[11] Method of exhaustion is a method of finding the area of a shape by inscribing inside it a sequence of polygons whose areas converge to the area of the containing shape. If the sequence is correctly constructed, the difference in area between the *n*th polygon and the containing shape will become arbitrarily small as *n* becomes large.

of their radii, volumes of spheres are to one another as the cubes of their radii, the volume of a pyramid is one-third the volume of a prism with the same base and height, and the volume of a cone is one-third that of the corresponding cylinder.

Eudoxus is considered to be the greatest of classical Greek mathematician, and in all antiquity second only to Archimedes (287–212 BC).

4.5 Euclid (c. 330–c. 260 BC)

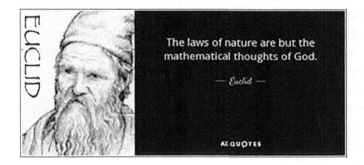

Euclid (Ref 14), also called Euclid of Alexandria to distinguish him from Euclid of Megara, was a Greek mathematician. He is considered the "Father of Geometry" and best known for his textbook *Elements* and the method for generating Pythagorean triples. More information on Pythagorean triples is provided in Appendix 8.

Elements is one of the most influential works in the history of mathematics, serving as the main textbook for teaching mathematics (especially geometry) from the time of its publication until the late 19th or 20th century. In the *Elements* Euclid deduced the principles of what is now called as Euclidean geometry from a small set of axioms.

Euclid's *Elements* (Ref 15) is a mathematical and geometric treatise containing of 13 books. It is a collection of definitions, axioms, theorems and constructions, and mathematical proofs of the theorems. The thirteen books cover Euclidean geometry and the ancient Greek version of elementary number theory. The work also includes an algebraic system that has become known as geometric algebra, which is powerful enough to solve many algebraic problems, including the problem of finding the square root of a number.

The *Elements* is the second oldest extant Greek mathematical treatises. The first oldest is *On the Moving Sphere* by Autolycus[12] and is the oldest extant axiomatic deductive treatment of mathematics. According to Proclus[13] the term "element" was used to describe a theorem that is all-pervading and helps furnishing proofs of many <u>other theorems.</u>

The *Elements* is a masterpiece in the application of logic to mathematics. It has proven enormously influential in many areas of science. Nicolaus Copernicus, Johannes Kepler, Galileo Galilei, and Sir Isaac Newton were all influenced by the *Elements* and applied their knowledge of it to their work. Albert Einstein recalled a copy of the *Elements* and a magnetic compass as two gifts that had a great influence on him as a boy, referring to the Euclid as the "holy little geometry book."

Euclidean geometry (Ref 16) is an axiomatic system, in which all theorems are derived from a small number of axioms. At the beginning of the first book of the *Elements*, Euclid gives five postulates (axioms) for plane geometry, stated in terms of constructions, as translated into English by Sir Thomas Heath[14].

"Let the following be postulated":

1. "To draw a straight line from any point to any point."
2. "To produce [extend] a finite straight line continuously in a straight line."
3. "To describe a circle with any center and distance [radius]."
4. "That all right angles are equal to one another."

---

[12] Autolycus of Pitane (c. 360–c. 290 BC) was a Greek astronomer, mathematician, and geographer.

[13] Proclus (412–485 AD) was a Greek philosopher, one of the last major Classical philosophers.

[14] Sir Thomas Heath (1861–1940) was a British mathematician, classical scholar, historian of ancient Greek mathematics, translator, and mountaineer. He also translated works by Apollonius of Perga (c. 262–c. 190 BC), by Aristarchus of Samos (c. 310–c. 230 BC), and by Archimedes of Syracuse (c. 287–c. 212 BC).

5. "That, if a straight line falling on two straight lines make the interior angles on the same side less than two right angles, the two straight lines, if produced indefinitely, <u>meet on that side on which are the angles less than the two right angles.</u>"

The fifth postulate is illustrated by a sketch shown below. Angles α and β are not the right angles, so the two straight lines are not parallel.

The *Elements* also include the following five "common notations":

1. Things that are equal to the same thing are also equal to one another.
2. If equals are added to equals, then the wholes are equal.
3. If equals are subtracted from equals, then the remainders are equal.
4. Things that coincide with one another are equal to one another.
5. The whole is greater than the part.

Among many other mathematical "pearls", the thirteen volumes of the *Elements* contain a proof that there are an infinite number of Pythagorean triples and the formulas for generating them are:

$a = m^2 - n^2$
$b = 2m \times n$
$c = m^2 + n^2$
where $m > n$ and $n > 0$

A Pythagorean triple consists of three positive integers *a*, *b*, and *c*, such as:

$$a^2 + b^2 = c^2.$$

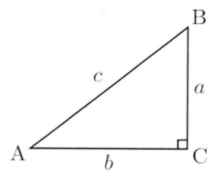

Such a triple is commonly written (*a*, *b*, *c*), is a Pythagorean triple. A triangle whose sides form a Pythagorean triple is called a Pythagorean triangle, and is necessary a right triangle. More detailed information on Pythagorean triples and the formulas for generating them is provided in Appendix 8.

In addition to geometry, Euclid also worked on algebra, solid geometry, spherical geometry, conic sections, number theory, optics, and mechanics, but only half of his writing survived.

As for the number theory in the time of ancient Greece, it was called the highest arithmetic—a branch of mathematics that originally studied the properties of integers. Euclid was the first who discovered perfect numbers: 6, 28, 496, and 8128.

4.6 Archimedes (c. 287–212 BC)

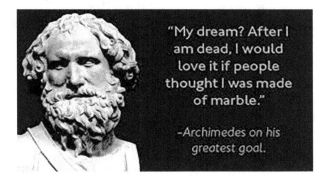

Archimedes of Syracuse (Ref 17) was a Greek mathematician, physicist, engineer, inventor, and astronomer. He is regarded as one of the leading scientists in classical antiquity. Generally considered greatest mathematician of antiquity and one of the greatest of all time. Archimedes used methods of infinitesimals and exhaustion in a way that is similar to modern integral calculus.

Archimedes applied the method of exhaustion to approximate the value of pi (π). He did this by drawing a larger regular hexagon outside a circle and a smaller regular hexagon inside the circle. Progressively doubling the number of sides of each regular polygon, he calculated the length of a side of each polygon at each step. As the number of sides increases, it becomes a more accurate approximation of a circle. After four such steps, when the polygons had 96 sides each, Archimedes was able to determine that the value of π lay between $3\frac{1}{7}$ (approximately 3.1429) and $3\frac{10}{71}$ (approximately 3.1408), consistent with its actual value of approximately 3.1416.

The calculations below are performed by using the Keisan Online Calculator for regular polygons inscribed to a circle (Ref 18). As an example, we chose a 96-side regular polygon (n = 96) inscribed to a circle (r = 1 unit of measurement: 1 in, 1 ft, 1 cm, or 1 m).

The calculated parameters are:

1. Polygon side length (a) $a \approx 0.06544$
2. Polygon perimeter (L) $L = a \times n = 0.06544 \times 96 = 6.28224$
3. Polygon area (S_p) $SP \approx 3.139350$
4. Circle circumference (C) $C = 2\pi r \approx 6.28319$
5. Circle area (S_c) $S_c = \pi r^2 \approx 3.14159$
6. Perimeter (L) to circumference (C) ratio:
   $L/C = 6.28224/6.28319 = 0.99985$
7. Polygon area (SP) to circle area (S_c) ratio:
   $S_p/S_c = 3.139350/3.14159 = 0.99929$

Difference between the inscribed circle circumference (C) and the perimeter of the 96-side regular polygon (L) is:

$6.28319 - 6.28224 = 0.00095$,

that is only $(0.00095/6.28319) \times 100\% \approx 0.015\%$

Difference between the inscribed circle area (S_c) and the polygon area (S_p) is:

$3.14159 - 3.13935 = 0.00224$,

that is only $(0.00224/6.28319) \times 100\% \approx 0.035\%$

Using the method of exhaustion, Archimedes defined the quantity of his "pi" that was less than the quantity of the modern π only by 0.015, or 0.035%

Those who are interested in calculating a side length of a regular polygon and its area, can use the following formulas (Ref 18):

Polygon side (a): $a = 2r \times \sin(\pi/n)$, where n is the number of sides,

Polygon area (S_p): $S_p = 1/2\, n \times r^2 \sin(2\pi/n)$, where r is a radius of a circle in which a polygon is inscribed.

Archimedes obtained the result of which he was most proud: the relationship between a sphere and a circumscribed cylinder of the same height and diameter (shown below). The volume is $4/3\, \pi r^3$ for the sphere, and $2\pi r^3$ for the cylinder. The surface area is $4\pi r^2$ for the sphere, and $6\pi r^2$ for the cylinder, including its two bases.

A sculpted ball and cylinder were placed on the tomb of Archimedes at his request.

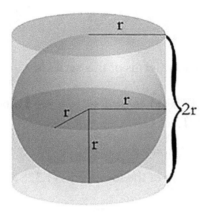

In "The Quadrature of the Parabola" Archimedes proved that the area enclosed by a parabola and a straight line is 4/3 times the area of a corresponding inscribed triangle as shown in this figure.

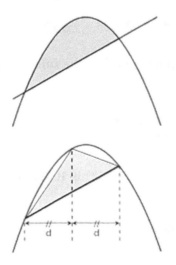

The area of the parabolic segment in the upper figure is equal to 4/3 that of thinscribed triangle in the lower figure. Archimedes expressed the solution to the problem as an infinite geometric series with the common ratio 1/4. In modern mathematical notation, it is:

$$\sum_{n=0}^{\infty} 4^{-n} = 1 + 4^{-1} + 4^{-2} + 4^{-3} + \cdots = \frac{4}{3}.$$

The sum of the four terms of this series is:

1 + 1/4 + 1/16 + 1/64 = 1 + (16 + 4 + 1)/64 = 1 + 21/64 = 1 + 0.328125 = <u>1.328125</u>
4/3 = <u>1.333333</u>

As you see, the sum of the four terms (1.328125) is less than 1.333333 by:

1.333333 − 1.328125 = <u>0.005208.</u>

Accepting that 4/3 (1.333333) is 100%, then the sum of the four terms is:

1.328125 × 100/1.333333 = <u>99.60%</u>

Let's add the fifth term (4^{-4}) to the sum of the four terms (1.328125):

4^{-4} + 1.328125 = 1/256 + 1.328125 = 0.003906 + 1.328125 = <u>1.332031</u>

Accepting that 4/3 (1.333333) is 100%, then the sum of the five terms is:

1.332031 × 100/1.333333 = <u>99.90%</u>

Let's add the sixth term (4^{-5}) to the sum of the five terms (1.332031):

$1.332031 + 4^{-5} = 1.332031 + 1/1024 = 1.332031 + 0.000977 = 1.333008$

Accepting that $4/3$ (1.333333) is 100%, then the sum of the six terms is:

$1.333008 \times 100/1.333333 = \underline{99.9756\%}$

Let's add the seventh term (4^{-6}) to the sum of the six terms (1.333008):

$1.333008 + 4^{-6} = 1.333008 + 1/4096 = 1.333008 + 0.000024 = 1.333032$

Accepting that $4/3$ (1.333333) is 100%, then the sum of the seven terms is:

$1.333032 \times 100/1.333333 = \underline{99.9774\%}$

Archimedes was one of the first to apply mathematics to hydrostatics and statics, including the explanation of the principle of the lever (work on levers caused him to remark: "Give me a place to stand on, and I will move the Earth").

Archimedes designed block-and-tackle pulley systems, allowing sailors to use the principle of leverage to lift objects that would otherwise have been too heavy to move.

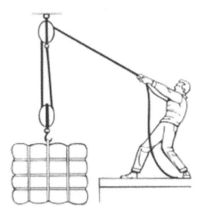

Archimedes studied the spiral known as the Archimedes screw. This screw was turned by manual labor. As the shaft turns, the bottom end scoops up a volume of water. This water is then pushed up the tube by rotating shaft until it pours out from the top of the tube.

The most widely known anecdote about Archimedes tells of how he invented a method for determining the volume of an object with an irregular shape. According to Vitruvius[15], a votive crown for a temple had been made of the pure gold, and Archimedes was asked to

---

[15] Marcus Vitruvius Pollio, commonly known as Vitruvius (c. 80–c. 15 BC), was a Roman architect, civil engineer, military engineer, and the author of *Ten Books on Architecture*.

determine whether some silver had been added into it by the dishonest goldsmith. Archimedes had to solve the problem without melting the <u>crown into a regularly shaped body in order to calculate its density.</u>

While taking a bath, Archimedes noticed that the level of the water in the tub rose as he got in and realized that this effect could be used to determine the volume of the crown. The submerged crown would displace an amount of water equal to its own volume. By dividing the mass of the crown by the volume of water displaced, the density of the crown could be obtained. This density would be lower than that of gold if less dense metals had been added. Archimedes then took to the streets naked, so excited by his discovery that he had forgotten to dress, crying "Eureka!" (Meaning "I have found it!"). The test was conducted successfully, proving that silver had indeed been mixed in.

Archimedes was killed in 212 BC during the Second Punic War (218–201 BC) by a Roman soldier after snapping at him "Don't disturb my circles," a reference to a geometric figure Archimedes had outlined on the sand.

4.7 Apollonius (c. 262–c. 190 BC)

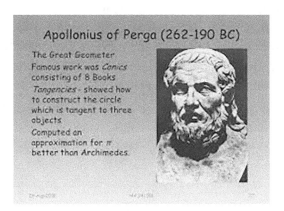

Apollonius of Perga (Ref 19) was a Greek mathematician and astronomer, best known for his writing on *Conic Sections*.

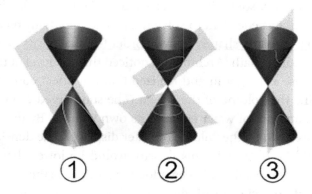

Types of conic sections: 1. Parabola, 2. Circle and Ellipse, 3. Hyperbola

His innovative methodology and terminology in the field of conics, influenced many later scholars, including Ptolemy (100–170 AD), Kepler (1571–1630), René Descartes (1596–1650), and Isaac Newton (1642–1726).

Apollonius coined the terminology that in use today for conic sections, namely *parabola*, *ellipse*, and *hyperbola*. His work *Conics* is one of the best known and preserved mathematical works from antiquity. He derives in it many theorems concerning conic sections that would prove invaluable to later mathematicians and astronomers studying planetary motion, such as Isaac Newton. Apollonius' treatment of curves is in some ways similar to the modern treatment, and some of his work seems to anticipate the development of analytic geometry by Descartes by some 1800 years later (Ref 2).

4.8 Hipparchus (c. 190–c. 120 BC)

Hipparchus of Nicaea (Ref 20) was a Greek astronomer, mathematician, and geographer. He is considered the founder of trigonometry for compiling the first known trigonometrical table, and the systematic use of the 360-degree circle.

Hipparchus is the most famous for his discovery of precession of the equinoxes[16]. <u>Axial precession is the movement of the rotational axis of Earth.</u>

---

[16] An equinox is commonly regarded as the instant of time when the plane (extended indefinitely in all directions) of Earth's equator passes through the center of the sun. This occurs twice each year: around 20 March and 23 September. In other words, it is the moment at which the center of the visible sun is directly above the equator.

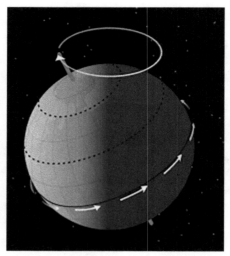

Earth rotates (white arrows) once a day around its rotational axis (red). This axis itself rotates slowly (white circle).

Hipparchus concluded that the equinoxes were moving (precessing) through the zodiac, and that the rate of precession was not less than 1° in 100 years.

According to modern astronomy, 1° of precession of the Earth takes 71.6 years. The axis of the Earth makes a complete turn (360°) in:

$$360° \times 71.6 = 25{,}776 \text{ years.}$$

"The climate on Earth (temperature, humidity, water level in the oceans, flora, fauna, and everything else) completely depends on the phenomenon of the precession of the Earth (the planet tilts toward the sun).

Geological knowledge says that the last ice age ended about 12,000 years ago. The ice shell reached the latitude of the cities of Berlin and London. Scandinavia, Baltic, Canada, and Siberia were under a multi-meter layer of ice and snow. Ice was the same as today in Antarctica. Due to the fact that the ice covered a large area of Eurasia and America—the water was concentrated in it, and from this the level of the world ocean was 150–200 meters lower than the modern one.

Now let's think…If the peak of cooling was 12 thousand years ago, then during the precession movement of the Earth's axis, the peak of heat in the northern hemisphere has not yet passed. The real heat (its peak) will be in a thousand years. Until then, there will be constant warming.

Now on the Earth for the northern hemisphere, in the years of precession is approximately the "beginning of summer." After the passage of the peak of heat on Earth by "inertia" for about four thousand years, there will be warming up.

We can predict the consequences. The modern deserts will expand, the steppes will dry up, and the zones of forests will turn into steppes. The swamps in the tundra will dry out and overgrow with deciduous forests. The Arctic Ocean will be moderately warm. Winters will be snowless. In the nearest thousand years all the ice will melt in the whole northern hemisphere, and from this, the level of the world ocean will rise. With the melting of the northern ice in the rays of the sun, the melted water will heat up. The warm "northern water" by the ocean currents will melt the ice of Antarctica and this will raise the total water level in the seas and oceans even more. As a result, large areas of modern coasts and cities will be under water.

No one today knows to what values the temperature on Earth will rise. Of course, you can simulate a situation, but humanity cannot change the precession climate change. The situation must be taken as well as the presence of the sun in the sky.

Such changes on Earth will occur every 26 thousand years. Humanity is still young in knowledge. In fact, it is in its "infancy," and it lives its first conscious life in the history of the Earth. Therefore, everything that happens on the Earth is the first lessons that teach humanity". This article was published in the Russian journal "Druzhba" ("Friendship") in April 2019. Translated into English by Edmund Isakov, the author of the book.

Hipparchus was recognized as the first mathematician known to have possessed a trigonometric table, which he needed when computing the eccentricity of the orbits of the moon and sun. He tabulated values for the chord function, which gives the length of the chord for each angle. He did this for a circle with a circumference of 21600 and a radius of 3438 units: this circle has a unit length of 1 arc minute along its perimeter. He tabulated the chords for angles with increments of 7.5°. In modern terms, the chord (*BX*) of an angle (Θ) equals the radius *r* (*MX* or *MB*) times twice the *sin* of half of the angle (Θ).

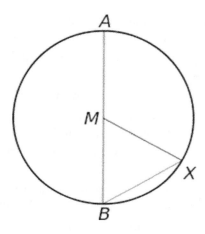

$$BX = 2\,r \times \sin\left(\frac{\Theta}{2}\right)$$

Pay attention to the numbers 21600 (circumference) and 3438 (the radius of the circle). Diameter of this circle is 6876. Dividing 21600 by 6876, we get 3.14136. Compare this number with pi (π) to five decimal places: 3.14159. The difference between them is only 0.00023, or 0.006%.

Not much is known about the life of Hipparchus, but he was apparently famous enough to have his face on Roman coins and the Greek post stamps. The Astronomer's Monument in Los Angeles, USA, features a relief of Hipparchus as one of six of the greatest astronomers of all time and the only one from antiquity.

4.9 Heron (c. 10–75 c. AD)

Hero of Alexandria (Ref 21), also known as Heron of Alexandria, was a Greek mathematician and engineer. He is considered the greatest experimenter of antiquity. Hero published a description of a steam-powered device called a "Hero engine," which was a rocket-like reaction engine. It was created almost two millennia before Industrial Revolution (transition to new manufacturing processes in the period from about 1760 to 1840).

Among Heron's most famous inventions was a windwheel, the earliest prototype of a windmill. A windwheel operating an organ was the first instant of wind powering a machine in history:

The first vending machine was also one of Heron's constructions. When a coin was inserted into a slot on the top of the machine, a set amount of a holy water was dispensing:

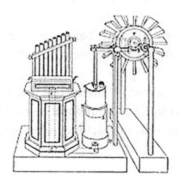

Much of Hero's original writings and designs have been lost, but some of his works were preserved in Arabic manuscripts.

As a mathematician, Heron described a method of iteratively computing the square root. Today, however, his name is most closely associated with "Heron's Formula" for finding the area of a triangle from its side lengths.

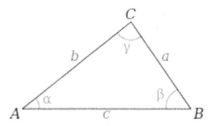

The area (A) of a triangle with sides a, b, and c is calculated by the formula:

$$A = \sqrt{s(s - a)(s - b)(s - c)},$$

where s is the semi-perimeter of the triangle. Semi-perimeter is calculated by the formula:

$$s = \frac{a + b + c}{2}$$

Let's △ ABC is a triangle with the sides: a = 4, b = 13, c = 15. The semi- perimeter s is:

$$s = 1/2\,(a + b + c) = 1/2\,(4 + 13 + 15) = 16$$

The area (A) is:

$$A = \sqrt{s(s - a)(s - b)(s - c)} = \sqrt{16 \cdot (16 - 4) \cdot (16 - 13) \cdot (16 - 15)}$$
$$= \sqrt{16 \cdot 12 \cdot 3 \cdot 1} = \sqrt{576} = 24.$$

In this example, the side lengths and area are all integers[17], making it a Heronian triangle. However, Heron's formula works equally well in cases where one or all <u>these numbers are not integers.</u>

In geometry, a Heronian triangle is a triangle that has side lengths and area that are all integers. Heronian triangles are named after Heron

---

[17] An integer (from the Latin *integer* meaning "whole") is a number that can be written without a fractional component. For example, 2, 13, 0, and 85 are integers, while $1\frac{2}{3}$, 3.14, and $\sqrt{5}$ are not.

of Alexandria. The term is sometimes applied more widely to triangles whose sides and area are all rational numbers. All rational numbers are integers.

There are some properties of Heronian triangles:

- Any right-angled triangle whose side lengths are a Pythagorean triple is a Heronian triangle, as the side lengths of such a triangle are integers, and its area is also an integer.
- The area of a Heronian triangle is always divisible by 6, for example:

    Pythagorean triples are $a = 5, b = 12, c = 13$.
    The area of this triangle is

    $1/2 (a \times b) = 1/2 (5 \times 12) = 30. \ 30/6 = 5$

- The perimeter of a Heronian triangle is always an even number, for example:

    The perimeter of the above Pythagorean triples is
    $a + b + c = 5 + 12 + 13 = 30$

- There are no equilateral Heronian triangles.
- There are no Heronian triangles with a side length of either 1 or 2

Heron is regarded as the greatest engineer in the history of mankind. He was the first to invent automatic doors, automatic puppet theaters, a sales machine, a quick-firing self-loading crossbow, a steam turbine, automatic decorations, an instrument for measuring the length of roads (an ancient odometer), etc. He was the first to create programmable devices: a shaft with pins with a rope wound on it.

4.10 Ptolemy (c. 85–c. 165 AD)

Claudius Ptolemy (Ref 2) was a Greco-Egyptian astronomer, mathematician, and geographer.

Ptolemy's major work is the *Almagest*, which is a treatise in thirteen books. However, it was not its original name. The original Greek title translates as *The Mathematical Compilation*, but this title was soon replaced by another Greek title, which means *The Greatest Compilation*. That title was translated into Arabic as "Al-majisti" and from this, the title *Almagest* was given to the work when it was translated from Arabic to Latin.

The *Almagest* is a 2nd-century astronomical and mathematical treatise on the apparent motions of the stars and planetary paths. It is one of the most influential scientific texts of all time with its geocentric model (the universe where the Earth is the orbital center of all celestial bodies) accepted for about 1400 years from its origin.

Ptolemy is also known for his "Ptolemy's Theorem" (Ref 22), which he used to creating a trigonometric table that he applied to astronomy.

The theorem states that: *If a quadrilateral is inscribable in a circle, then the product of the measures of its diagonals is equal to the sum of the products of the measures of the pairs of opposite sides.*

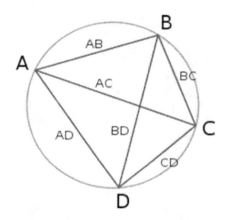

The theorem is expressed by the formula:

AC × BD = AB × CD + BC × AD

The converse of Ptolemy's theorem is also true,

In a quadrilateral, if the sum of the products of its two pairs of opposite sides is equal to the product of its diagonals, then the quadrilateral can be inscribed in a circle:

AB × CD + BC × AD = AC × BD

Claudius Ptolemy was one of the most influential Greek astronomers of his time. However, of all the ancient Greek mathematicians, it is fair to say that his work has generated more discussion and argument than any other.

Nicolaus Copernicus (1473–1543), astronomer and mathematician, formulated a model of the universe that placed the Sun rather than the Earth at the center of the universe. An ancient Greek Aristarchus of Samos (c. 310–c. 230 BC), astronomer and mathematician, had formulated such a model some eighteen centuries earlier.

4.11 Diophantus (c. 200–c. 284 AD)

Diophantus of Alexandria (Ref 23) was a Greek mathematician, sometimes called "the father of algebra." His major work was the *Arithmetica*, the most prominent work on algebra. Of the original thirteen books of which *Arithmetica* consisted, only six have survived. There are some who believe that four Arab books[18] are also by Diophantus. Some problems from *Arithmetica* have been found in Arabic sources.

Diophantus lived in Alexandria, Egypt. Much of our knowledge of his life is derived from a 5th-century Greek anthology of number games and puzzles. One of the puzzles (sometimes it is called the Diophantus epitaph) states:

---

[18] J. Sesiano (1982). *Books IV to VII of Diophantus'* "Arithmetica" *in the Arabic Translation Attributed to Qusta ibn Luqa*. New York/Heidelberg/Berlin: Springer-Vertag, p. 502.

Here lies Diophantus, the wonder behold.
Through art algebraic, the stone tells how old:
God gave him his boyhood one-sixth of his life,
One-twelfth more as youth while whiskers grew rife.
And then yet one-seventh ere marriage begun.
In five years there came a bouncing new son.
Alas, the dear child of master and sage
After attaining half the measure of his father's life
Chill fate took him. After consoling his fate by the science
Of numbers for four years, he ended his life.

In modern mathematical notation, this puzzle is defined as a linear equation with one unknown:

$$x = \frac{x}{6} + \frac{x}{12} + \frac{x}{6} 5 + \frac{x}{2} + 4$$

Let x be the number of years Diophantus lived. The equation is simplified to:

$$x - 9 = \frac{x}{6} + \frac{x}{12} + \frac{x}{7} + \frac{x}{2}$$

$$x - 9 = \frac{14x + 7x + 12x + 42x}{84}$$

$84x - 84 \times 9 = 14x + 7x + 12x + 42x$
$84x - 756 = 75x$
$84x - 75x = 756$
$9x = 756$
$x = 756/9 = 84$

Diophantus lived 84 years.
Diophantus recognized fractions as numbers. He allowed positive ration numbers for coefficients and solutions.
In modern use, Diophantine equations are usually algebraic equations with integer coefficients, for which integer solutions are sought.
Diophantus' work created a foundation for work on algebra and in fact much of advanced mathematics is based on algebra.

Diophantus is often called "the father of algebra" because he contributed greatly to number theory, mathematical notation, and arithmetic.

4.12 Hypatia of Alexandria (c. 370–415 AD)

Hypatia (Ref 24) was a Greek mathematician, astronomer, and philosopher in Egypt. At the time, Egypt was a part of the Byzantine Empire. Hypatia was educated in Athens. Around 400 AD, she became head of the Neoplatonist School[19] in Alexandria, where she imparted the knowledge of Plato and Aristotle and taught <u>philosophy and astronomy to students, including pagans, Christians, and foreigners.</u>

---

[19] Neoplatonism is a modern term used to designate a tradition of philosophy that arose in the 3rd century AD and persisted until shortly after the closing of the Platonic Academy in Athens in 529 AD by Justinian I, a Byzantine (East Roman) emperor from 527 to 565. Neoplatonists were heavily influenced by Plato, but also by the Platonic tradition that thrived during the six centuries.

No written work widely recognized by scholars as Hypatia's own has survived to the present time. Many of the works commonly attributed to her had been collaborative works with her father, Theon of Alexandria. This kind of authorial uncertainty is typical for female philosophers in antiquity.

A partial list of Hypatia's works as mentioned by other antique and medieval authors or as posited by modern authors:

- A commentary on the 13-volume *Arithmetica* by Diophantus.
- A commentary on the *Conics* of Apollonius of Perga.
- Edited the existing version of Ptolemy's *Almagest*. "Until recently scholars thought that Hypatia revised Theon's commentary on *Almagest*. The view was based on the title of the commentary on the third book of *Almagest*, which read: "Commentary by Theon of Alexandria on Book III of Ptolemy's *Almagest*, edition revised by my daughter Hypatia, the philosopher." Cameron, who analyzed Theon's titles for other books of *Almagest* and for other scholarly texts of late antiquity, concludes that Hypatia corrected not her father's commentary but the text of *Almagest* itself. Thus, the extant text of *Almagest* could have been prepared, at least partly, by Hypatia".
- Edited her father's commentary on Euclid's *Elements*.
- She wrote a text *"The Astronomical Canon"* (a new edition of Ptolemy's *Handy Tables* or commentary on the prior *Almagest*).

Hypatia contributions to technology are supposed to include the invention of the hydrometer used to determine the specific gravity[20] of liquids. However, the hydrometer was invented before Hypatia, and already known in her time. Some say that this is a textual misinterpretation of the original Greek, which mentions a hydroscopium—a

---

[20] Specific gravity is the ratio of the weight of a volume of the substance to the weight of an equal volume of the reference substance. The reference substance is nearly always water at its densest (4° C) for liquids.

clock that works with water and gears, similar to the Antikythera[21] mechanism.

Hypatia was murdered during a city-wide anger stemming from a feud between Orestes[22], the prefect of Alexandria, and Cyril[23], the Bishop of Alexandria.

---

[21] Antikythera mechanism is an ancient analogue computer used to predict eclipses and the cycles of the ancient Olympic Games.
[22] Orestes was the Roman governor of the province of Egypt in 415.
[23] Cyril of Alexandria was the Patriarch of Alexandria from 412 to 444.

CHAPTER 5

CHINESE MATHEMATICS

Early Chinese mathematics (Ref 2, 12) is different from that of other parts of the world and was developing independently. The older extant mathematical text from China dated to between 1200 BC and 300 BC.

The simple but efficient ancient Chinese numbering system, which dates back to at least the 2nd millennium BC, used small bamboo rods arranged to represent the numbers 1 to 9, which were then places in columns representing units, tens, hundreds, thousands, etc. It was therefore a decimal place value system, very similar to the one we use today—indeed it was the first such number system, adopted by the Chinese over a thousand years before it was adopted in the West.

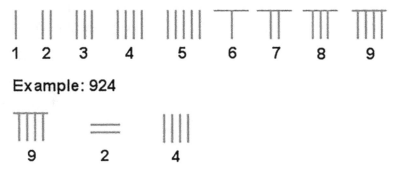

Ancient Chinese number system

Rod numerals allowed the representation of numbers as large as desired and to perform calculations by using the *suan pan*, or Chinese abacus.

The date of the invention of the *suan pan* is not certain, but the earliest written mention dates from 190 BC.

The oldest existent work on geometry in China comes from the philosophical Mohist[1]) canon around 330 BC, compiled by the followers of Mozi (470–391 BC).

In 212 BC, Emperor Qin Shi Huang commanded all books in the Qin Empire other than officially sanctioned ones be burned. This decree was not universally obeyed, but as a consequence of this order, little is known about ancient Chinese mathematics before this date.

After the book burning of 212 BC, the Han dynasty (202 BC–220 AD) produced works of mathematics, which presumably expanded on works that are now lost.

The main thrust of Chinese mathematics developed in response to the empire's growing need for mathematically competent administrators. A textbook called "Jiuzhang Suanshu" or "Nine Chapters on the Mathematical Art" (written over a period of time from about 200 BC onwards, probably by a variety of authors) became an important tool in the education of such a civil service, covering hundreds of problems

---

[1] Mohism, also known as Mohist School of Logic, literally: "School of Mo", was an ancient Chinese philosophy of logic, rational thought and science developed by the academic scholars who studied under the ancient Chinese philosopher Mozi.

in practical areas such as trade, taxation, engineering and the payment of wages.

By the 13th century, the Golden Age of Chinese mathematics, there were over 30 prestigious mathematics schools scattered across China.

5.1 Liu Xin (c. 50 BC–23 AD)

Liu Xin (Ref 25) was an ancient Chinese astronomer, mathematician, historian, librarian, and editor.

As a curator of the imperial library, Liu Xin was the first to establish a library classification system and the first book notation system. At his time, the library catalog was written on scrolls of fine silk and stored in silk bags. He catalogued and annotated or edited ancient texts. His projects produced what became definitive texts of a number of orthodox canons of Chinese philosophy and history.

The Chinese for centuries had used the value of 3 for their calculation of the ratio of a circle's circumference to its diameter (in modern mathematics, it is π). Between the years 1 and 5 AD, Liu Xin was the first to give a geometrical figure, which implies a more accurate value of π at 3.1457 (the exact method he used to reach this figure is unknown). Jialiang is an ancient Chinese measuring device for several volume standards that were calculated using the value of π at 3.1457.

5.2 Zhang Heng (78–139 AD)

Zhang Heng (Ref 26) was an ancient Chinese astronomer, mathematician, scientist, engineer, inventor, geographer, cartographer, statesman, artist, poet, and literary scholar. He studied the relationship between celestial circle and diameter of the earth. In his measurements, the celestial circle was 736 units of length and the diameter of the earth was 232 units of length. Thus, the ratio (R) of the celestial circle to diameter of the earth is:

R = 736/232 = 3.1724

At Heng's time, the ratio 4:3 was given for the area of a square to the area of its inscribed circle, as shown in the figure below.

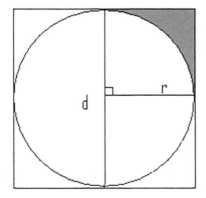

Let's compare the areas of the inscribed circle counted by Zhang Heng and counted by the generally accepted formula.

For example, diameter of the inscribed circle and the side of a square are $d = 2$. Radius of the circle is $r = 1$. In modern notation, the area of a circle (A_C) is:

$$A_C = \pi r^2 = \pi \times 1^2 = 3.14159$$

The area of a square (A_S) is:

$$A_S = d^2 = 2^2 = 4$$

Ratio of the area of a square to the area of the inscribed circle is:

$$A_S/A_C = 4/3.14159 = 1.27324$$

At Heng's time, the ratio of a square to the area of the inscribed circle (R_H) was:

$$R_H = 4/3 = 1.33333$$

Actual ratio of the area of a square to the area of the inscribed circle (R_A) is:

$$R_A = 1.27324$$

In this example, $R_H = 1.33333$ and $R_A = 1.27324$, so R_H is greater than R_A by:

$$1.33333 - 1.27324 = 0.06009$$

Expressing the value of 0.06009 as a percentage, we get:

$$0.06009 \times 100/1.27324 = 6.006/1.27324 = 4.7\%$$

By the way, constant ratio of the circumference to diameter of any circle is old as man's desire to measure. The symbol for this ratio known as π (pi) dates from 1706. It was first used in print by William Jones (1675–1749), a Welsh mathematician.

Zhang Heng considered that the volume of a cube and volume of the inscribed sphere should be $4^2 : 3^2$.

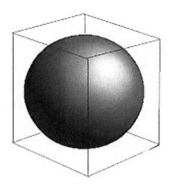

In formula, with D is a diameter and V is a volume, $D^3 : V = 16:9$ or $V = 9/16\ D^3$. Zhang Heng realized that the value for diameter in this formula was inaccurate, noting the discrepancy as the value taken for the ratio. He then attempted to remedy this by amending the formula with an additional $1/16\ D^3$, hence:

$$V = 9/16\ D^3 + 1/16\ D^3 = 5/8\ D^3$$

Let's see, which one of the two formulas for the volume of a sphere inscribed in cube, is more accurate. The original formula is:

$$V_1 = 9/16\ D^3 \qquad (1)$$

The amended formula is:

$$V_2 = 5/8\ D^3 \qquad (2)$$

Diameter of a sphere inscribed in a cube, and sides of a cube are of the same size.

For example, diameter of a sphere and sides of a cube are equal 4. The volume of a sphere calculated by formula (1):

$$V_1 = 9/16\ D^3 = (9/16) \times 4^3 = 0.5625 \times 64 = \underline{36} \qquad (1)$$

The volume of a sphere calculated by formula (2):

$$V_2 = 5/8\, D^3 = (5/8) \times 4^3 = 0.625 \times 64 = \underline{40} \qquad (2)$$

The volume of a cube (V_C) with sides equaled to diameter of inscribed sphere is:

$$V_C = D^3 = 4^3 = 64$$

In modern mathematical notation, volume of a sphere is calculated by formula (3)

$$V_S = \tfrac{4}{3}\pi r^3$$

$$V_S = \tfrac{4}{3}\pi 2^3 = 1.33333 \times 3.14159 \times 8 = \underline{33.51} \qquad (3)$$

Where r is the radius of the sphere (r = D/2)

As can be seen, formula (1) is more accurate than formula (2) in comparison with formula (3), which is in use worldwide.

In this example, $V_1 = 36$ and $V_S = 33.51$, so V_1 is greater than V_S by:

$[(V_1/V_S \times 100) - 100] =$
$[(36/33.51 \times 100) - 100] =$
$[107.4306 - 100] \approx 7.4\%$

Take into consideration that Zhang Heng developed formula (1) about 1900 years ago, so 7% difference is a good accuracy at that time. This formula could be more accurate, if he subtracted $1/16\, D^3$ from $9/16\, D^3$, rather than adding this value. It would be:

$$V_1 = 9/16\, D^3 - 1/16\, D^3 = 8/16\, D^3 = 1/2\, D^3 \qquad (4)$$

The volume of a sphere calculated by formula (4) is:

$$1/2\, D^3 = (1/2) \times 4^3 = 0.5 \times 64 = 32$$

If V1 = 32 and VS = 33.51, therefore V1 is less than VS by:

$[(V_1/V_S \times 100) - 100] =$
$[(32/33.51 \times 100) - 100] =$
$[95.49389 - 100] \approx -4.5\%$

Zhang Heng used division of 730 by 232 for calculating the volumes of spheres:

730/232 = 3.14655. Number of π with 5 decimal digits is 3.14159. The difference between the Zhang Heng's number of 3.14655 and number of π is:

3.14655 − 3.14159 = 0.00496, or 0.16%

As we can see, the number 3.14655 is very close to number π and it was known to Chinese mathematicians at the 2nd century AD.

5.3 Liu Hui (c. 225–c. 295 AD)

Liu Hui (Ref 27) was an ancient Chinese mathematician. In 263, he edited and published a book with solutions to mathematical problems presented in the famous Chinese book of mathematics known as *The Nine Chapters on the Mathematical Art*.

He was known as one of the greatest mathematicians of ancient China. He expressed all of his mathematical results in the form of decimal fractions using all aspects of measurements. Liu provided commentary on a mathematical proof of a theorem identical to the Pythagorean Theorem. Liu called the figure of the drawn diagram for the theorem the "diagram giving the relations between the hypotenuse and the sum and difference of the other two sides whereby one can find the unknown from the known." In the field of plane areas and solid figures, Liu was one of the greatest contributors to empirical solid geometry.

Liu found that a wedge with rectangular base and both sides sloping could be broken down into a pyramid and a tetrahedral wedge.

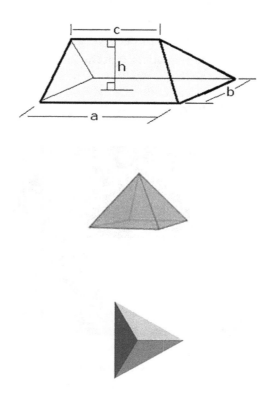

Liu Hui (Ref 28) was the first Chinese mathematician, who noted in his *The Nine Chapters on the Mathematical Art* that the ratio of the perimeter of an inscribed hexagon to the diameter of the circle was 3. So the ratio of the circumference of such circle to its diameter (π in modern mathematical notation) must be greater than 3. He suggested that 3.14 was a good enough approximation and expressed it by the improper fraction as 157/50 = 3.14.

Liu Hui proved an inequality of π by considering the area of inscribed polygons with N and $2N$ sides.

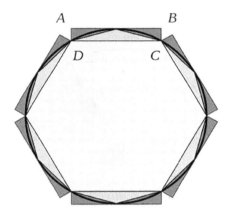

In the above figure, the yellow polygon represents the area of N-side polygon, denoted by A_N, and the yellow area plus the green area represents the area of $2N$-side polygon, denoted by A_{2N}.

The green area represents the difference ($D2N$) between the areas of the $2N$-sided polygon and the N-sided polygon:

$$D_{2N} = A_{2N} - A_N$$

The red area is equal to the green area, and so is also D_{2N}. Therefore,

Yellow area + green area + red area = $A_{2N} + D_{2N}$

Let C represent the area of the circle. Then

$$A_{2N} < C < A_{2N} + D_{2N}$$

If the radius of the circle equals 1 unit, then we have Liu Hui's π inequality:

$$A_{2N} < \pi < A_{2N} + D_{2N}$$

Liu Hui developed an iterative algorithm to calculate π to any required accuracy based on bisecting polygons. He began with an inscribed hexagon.

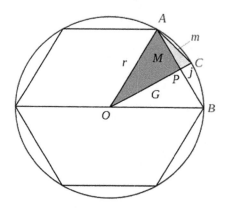

Let a side AB of hexagon equals M, and r is the radius OA of a circle. Bisect AB with line OPC, AC becomes a side of dodecagon. Let the length of AC be m. APO, APC are two right angle triangles. Liu Hui used Pythagorean Theorem repetitively:

$$G^2 = r^2 - \left(\frac{M}{2}\right)^2$$

$$G = \sqrt{r^2 - \frac{M^2}{4}}$$

$$j = r - G = r - \sqrt{r^2 - \frac{M^2}{4}}$$

$$m^2 = \left(\frac{M}{2}\right)^2 + j^2$$

$$m = \sqrt{m^2}.$$

With $r = 10$ units, he calculated areas (A) of 96-sided and 192-sided polygons:

96-sided polygon, $A96 = 313\frac{584}{625}$

192-sided polygon, $A192 = 314\frac{64}{625}$

Difference of 192-sided polygon and 96-sided polygon is:

$D_{192} = 314\frac{64}{625} - 313\frac{584}{625} = \frac{105}{625} = 0.168$

The calculations below, we were performed by using the "Keisan Online Calculator" for regular polygons inscribed to a circle (Ref 18), to compare them with the calculations performed by Liu Hui.

The 96-sided and 192-sided regular polygons inscribed to a circle are used to calculate the side length and the area of each polygon.

The side length of a polygon (a) is calculated by the formula:

$a = 2r\sin(\pi/n)$

Where r is the radius of a circle circumscribes a polygon

The area of a polygon (A) is calculated by the formula:

$A = 0.5nr^2 \sin(2\pi/n)$

Where n is the number of sides

Let the radius of the circle be the following: $r = 10$ unit (10 in, or 10 ft, or 10 cm, or 10 m).

Calculated values

96-sided polygon		192-sided polygon	
Side length	0.6543816	Side length	0.3272346
Area	313.935020	Area	314.103195

Comparison of the areas calculated by Liu Hui and by the author of this book

96-sided polygon	
Liu Hui	Author
$313\frac{584}{625}$ 313.93$\underline{44}$	313.93$\underline{50}$

192-sided polygon	
Liu Hui	Author
$314\frac{64}{625}$ 314.10$\underline{24}$	314.10$\underline{32}$

The difference between the values of the polygon areas calculated by Liu and by today's calculator is only on the two last decimal digits of the above values.

From Liu Hui's π inequality:

$$A_{2N} < C < A_{2N} + D_{2N}$$

$$A_{192} < C < A_{192} + D_{192}$$

$$314\frac{64}{625} < C < 314\frac{64}{625} + \frac{105}{625}$$

$$314\frac{64}{625} < C < 314\frac{169}{625}$$

In modern mathematical notation, C is calculated by the formula:

$C = \pi r^2$

Since $r = 10$, $C = \pi r^2 = \pi \times 100$, therefore:

$314\frac{64}{625} < \pi \times 100 < 314\frac{169}{625}$

In modern mathematical notation, the Liu Hui's π inequality is:

$3.141024 < \pi < 3.142704$

Note: π with the first 6 decimal places is 3.141592.

Liu Hui never took π as the average of the lower limit 3.141024 and upper limit 3.142704.

5.4 Zu Chongzhi (429–500 AD)

Zu Chongzhi (Ref 29, 30) was a Chinese mathematician, astronomer, writer, and politician. The majority of his great mathematical works were recorded in the lost text the *Zhui Shu*. Most scholars argue about his complexity since traditionally the Chinese had developed mathematics as algebraic and equational.

Zu's family had been involved in astronomy research, and from childhood Zu was taught astronomy and mathematics. When Emperor of the Chinese dynasty Liu Song, Liu Jun (430–464), heard of Zu, he was sent to an Academy and later at the Imperial Nanjing University to carry out research in mathematics.

Zu Chongzhi, along with his son Zu Gengzhi wrote a mathematical text entitled *Zhui Shu (Methods for Interpolation)*. It is said that the treatise contains formulas for the volume of the sphere, cubic equations, and the value of π. This book has been lost since the Song Dynasty.

Zu Chongzhi's works on the accurate value of the involved the lengthy calculations. He used the Liu Hui's value of π algorithm to inscribe a 12,288-sided polygon. Chongzhi's best approximation of π with the fraction 355/113 (decimal fraction is: 3.1415929). Zu's value of π is precise to six decimal places. For a thousand years thereafter no subsequent mathematician could compute π value with such precise.

In Chinese literature, the fraction 355/113 is known as "Zu's ratio". Zu's ratio is a best Zu's value of π is precise to six decimal places. For a thousand year thereafter no subsequent mathematician approximation to π.

Zu Chongzhi also worked on deriving the formula for the volume of a sphere:

$$V = \frac{355}{113} \times D^3/6 \qquad (1)$$

where V is a volume and D is a diameter.

The following formula is used for calculating the volume of a sphere in the modern mathematical notation:

$$V = 4\pi r^3/3 \qquad (2)$$

where r is the radius of a sphere.

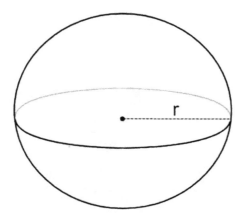

Let's calculate the volume of a sphere with the diameter of 8 units (no matter if it is 8 inches or 8 centimeters) using formula (1), and the same sphere with the radius of 4 units (4 inches or 4 centimeters) using formula (2).

Volume of the sphere calculated by formula (1):

$$V = \frac{355}{113} \times D^3/6 = \frac{355}{113} \times 8^3/6 = \frac{355}{113} \times 512/6 = \frac{355}{113} \times 85.333333 = 268.0825\underline{958}$$

Volume of the sphere calculated by formula (2):

$$V = 4\pi r^3/3 = 4\pi \times 4^3/3 = 4\pi \times 64/3 = 21.33 = 268.0825\underline{731}$$

As you can see, the formula Zu Chongzhi (1), which was used more than 1500 years ago, is almost as accurate as the formula (2), which we use today.

5.5 Yang Hui (1238–1298)

Yang Hui (Ref 31) was a Chinese mathematician. Yang worked on magic squares, magic circles, the binomial theorem, and is best known for his contribution of presenting Yang Hui's Triangle[2]).

Around 1275, Yang had two published mathematical books, in which he wrote on arrangement of natural numbers around concentric and non-concentric circles, and vertical-horizontal diagrams of complex combinatorial arrangements known as magic squares and magic circles. The books provided rules for their construction.

Magic square (Ref 32) is an arrangement of distinct numbers (i.e., each number is used once), usually integers, in a square grid, where the numbers in each row, in each column, and the numbers in the diagonals, all add up to the same number.

---

[2] In modern mathematics, it is known as Pascal's Triangle, which is a triangular array of the binomial coefficients. It is named after French mathematician Blaise Pascal (1623–1662). Other mathematicians studied a triangular array in India, Persia (Iran), China, Germany, and Italy centuries before Pascal.

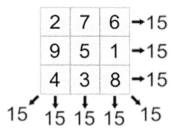

The constant that is the sum of every row, column and diagonal is called the magic constant or magic sum, M. Every normal magic square has a constant dependent on n, calculated by the formula:

$$M = [n(n^2 + 1)]/2$$

For normal magic squares of order n = 3, 4, 5, 6, 7, and 8, the magic constants are, respectively: 15, 34, 65, 111, 175, and 260.

For example:

Let's $n = 4$, then $M = [4 \times (4^2 + 1)]/2 = (4 \times 17)/2 = 68/2 = 34$

16	3	2	13
5	10	11	8
9	6	7	12
4	15	14	1

Magic squares have a long history, dating back to 650 BC in China. Magic circles were invented by Yang Hui, one of them is shown below (Ref 33).

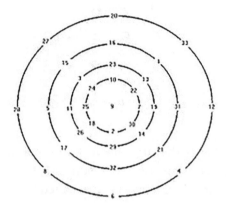

It is the arrangement of natural numbers on circles where the sum of the numbers on each circle and the sum of numbers on diameter are identical. This magic circle was constructed from 33 natural numbers from 1 to 33 arranged on four concentric circles, with 9 at the center.

CHAPTER 6

INDIAN MATHEMATICS

The earliest civilization (Ref 2) on the Indian subcontinent in the Indus Valley Civilization that flourished between 2600 and 1900 BC in the Indus River basin. Their cities were laid out with geometric regularity, but none of mathematical documents survived from this civilization.

The Hindu-Arabic numerals were invented by mathematicians in India and were called "Hindu numerals." Later they were called "Arabic numerals" by Europeans because they were introduced in the West by Arab merchants.

Various symbol sets are used to represent numbers in the Hindu-Arabic numeral system, all of which evolved from the Brahmi numerals. Each of the dozens major scripts of India has its own numeral glyphs. As an example, the following table shows Brahmi numerals (lower row) in India in the 1st century AD.

1	2	3	4	5	6	7	8	9
−	=	≡	+	h	ᛞ	?	ᔕ	?

The oldest mathematical records, dated between the 8th century BC and the 2nd century AD, give methods for constructing a circle with approximately the same area as given square, which imply several different approximations of the value of pi.

Classical period (Ref 34) of Indian mathematics (400–1600 AD) is known as the golden age. At that time, mathematicians gave broader and clear shape to many branches of mathematics. Their contributions spread to Asia, the Middle East, and eventually to Europe.

6.1 Aryabhata (476–550 AD)

Aryabhata (AD 476 – 550)

Aryabhata (Ref 35) was the first of the major mathematician-astronomers of the classical period. His astronomical treatise was written when he was 23 years old. It is only known surviving work of the 5th century.

Aryabhata's major work, *Aryabhatiya*, a brief summary of mathematics and astronomy, has survived to modern times. The mathematical part of the *Aryabhatiya* covers arithmetic, algebra, plane and spherical trigonometry, continued fractions, quadratic equations, and a table of sines.

Aryabhata worked on the approximation for pi and have conclude that pi is irrational. He writes: "Add 4 to 100, multiply by 8, and then add 62,000. By this rule the circumference of a circle with a diameter of 20000 can be approached." This implies that the ratio of the circumference to the diameter is:

[(4 + 100) × 8 + 62000]/20000 = 3.1416, which is correct to three decimal places.

It is speculated that Aryabhata used the word *approaching*, to mean that not only is this an approximation, but that the value is incommensurable (or irrational). If this is correct, it is quite a sophisticated insight, because the irrationality of pi was proved in Europe only in 1761 by Johann Heinrich Lambert (1728–1777), a Swiss polymath.

In *Aryabhatiya*, Aryabhata provided his formulas for summation of series of squares and cubes (a series is the sum of the terms of an infinite sequence, the sum of a finite sequence has defined first and last terms):

$$1^2 + 2^2 + \cdots + n^2 = \frac{n(n+1)(2n+1)}{6}$$

$$1^3 + 2^3 + \cdots + n^3 = (1 + 2 + \cdots + n)^2$$

Let's calculate the sum of 5 terms (Σ) of series of squares. The left-hand part of the equation is:

Σ = 1 + 4 + 9 + 16 + 25 = 55

The right-hand part of the equation:

$$\frac{5(5+1)(2\times5+1)}{6} = \frac{30\times11}{6} = 55$$

Let's calculate the sum of 5 terms (Σ) of series of cubes. The left-hand part of the equation:

Σ = 1 + 8 + 27 + 64 + 125 = 225

The right-hand part of the equation:

$$(1 + 2 + 3 + 4 + 5)^2 = 15^2 = 225$$

Aryabhata's work was of great influence in the Indian astronomical tradition and influenced several neighboring cultures through translations. The Arabic translation during the Islamic Golden Age (about 820 AD), was particularly influential. Some of it results are cited by Al-Khwarizmi (his contribution to mathematics is provided in Chapter 7) and Al-Biruni (973–1050 AD) that Aryabhata's followers believed that the Earth is rotated about its axis.

6.2 Varahamihira (505–587 AD)

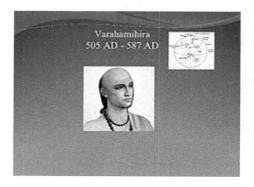

Varahamihira (Ref 36) also called Varaha or Mihir, was an Indian mathematician, astronomer, and astrologer. His mathematical work on trigonometry resulted in the development of these three formulas:

$$\sin^2 x + \cos^2 x = 1$$

$$\sin x = \cos\left(\frac{\pi}{2} - x\right)$$

$$\frac{1 - \cos 2x}{2} = \sin^2 x$$

The following calculations are taken to show that these formulas are accurate. The first formula (on the top):
Let's $x = 30°$

$$\sin^2 30° + \cos^2 30° = 0.5^2 + 0.866025404^2 = 0.25 + 0.75 = 1$$

The second formula (in the middle):
Let's $x = 30°$

$\sin 30° = 0.5$
$\cos (\pi/2 - 30°) = \cos (90° - 30°) = \cos 60° = 0.5$

The third formula (on the bottom):
Let's $x = 30°$

$\frac{1-\cos 60}{2} = \frac{1-0.5}{2} = 0.25$

$\sin^2 30° = 0.5^2 = 0.25$

Conclusion: all three formulas are correct

6.3 Brahmagupta (597–668 AD)

597 - 668 AD

Brahmagupta (Ref 37) was an astronomer and mathematician. He wrote many textbooks. The *Brahmasphutasiddhanta* meaning the *Corrected Treatise of Brahma* is one of his well-known works. Several chapters of this book were on mathematics.

One of his the most significant inputs to mathematics was the introduction of "zero" as a number, which means "nothing."

Brahmagupta was the first to give rules to compute with zero as a number. When zero is added to a number or subtracted from a number, the number remains unchanged. A number multiplied by zero becomes zero. Zero divided by any other number is zero. He believed that dividing zero by zero produces zero.

Mathematicians have now shown that zero divided by zero is undefined—it has no meaning. There really is no answer to zero divided by zero. Brahmagupta gave two equivalent formulas for solving the general quadratic equation, which in modern mathematical notation is expressed as follows:

$$ax^2 + bx + c = 0$$

Where x represents an unknown; a, b, and c represent known numbers such that $a \neq 0$ (is not equal 0), then the numbers a, b, and c are the coefficients of the equation.

First formula:

$$x = \frac{\sqrt{4ac + b^2} - b}{2a}$$

Second formula:

$$x = \frac{\sqrt{ac + \frac{b^2}{4}} - \frac{b}{2}}{a}.$$

Currently, we use general quadratic equation (shown below), which is equivalent to the First formula developed by Brahmagupta. Sign ± means that unknown x has two roots: x_1 and x_2

$$x = \frac{-b \pm \sqrt{b^2 - 4ac}}{2a}$$

As an example, let's solve the following quadratic equation:

$$x^2 + 3x - 4 = 0$$

$$x_1 = \frac{-3+\sqrt{3^2-4\times1\times(-4)}}{2\times1} = \frac{-3+\sqrt{25}}{2} = \frac{2}{2} = 1$$

$$x_1 = \frac{-3+\sqrt{3^2-4\times1\times(-4)}}{2\times1} = \frac{-3-\sqrt{25}}{2} = \frac{-8}{2} = -4$$

In geometry, Brahmagupta's theorem (Ref 38) states that if an inscribed quadrilateral is orthodiagonal (that is, has perpendicular diagonals), then the perpendicular to a side from the point of intersection of the diagonals always bisects the opposite side.

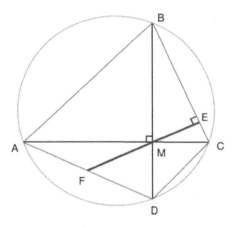

More specifically, let A, B, C, and D be four points on a circle such that the lines AC and BD are perpendiculars. Denote the intersection of AC and BD by M. Drop the perpendicular from M to the line BC, calling the intersection E. Let F be the intersection of the line EM and the edge AD. Then, the theorem states that F is the midpoint AD. We need to prove that AF = FD, then both AF and FD equal to FM.

To prove that AF = FM, note that the angles DAC and DBC are equal, because they are inscribed angles that intercept the same arc CD of the circle. Furthermore, the angles CBM and CME are both complementary to angle BCM (i.e., they add up to 90°), and, therefore, are equal. Finally, the angles CME and FMA are the same. Hence, AFM is an isosceles triangle, and thus the sides AF and FM are equal.

The proof that FD = FM goes similarly: the angles FDM, BCM, BME, and DMF are equal, so DFM is an isosceles triangle, so FD = FM. It follows that AF = FD, as the theorem claims.

Brahmagupta's most famous result in geometry is his formula for inscribed quadrilaterals. Given the lengths of the sides of any inscribed quadrilateral, Brahmagupta gave an approximate and an exact formula for the figure's area.

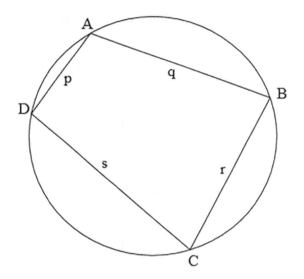

Let ABCD is the inscribed quadrilateral. AB = q, BC = r, CD = s, and AD = p. The approximate area (A_1) is:

$$A_1 = \frac{p+r}{2} \times \frac{q+s}{2}$$

The exact area (A_2) is:

$$A2 = \sqrt{(t-p)(t-q)(t-r)(t-s)}$$

where t is a semi-perimeter of the inscribed quadrilateral.

$$t = \frac{p+q+r+s}{2}$$

6.4 Bhāskara I (c. 600–c. 680 AD)

Bhāskara (Ref 39), commonly called Bhāskara I to avoid confusion with the 12th century mathematician Bhāskara II, was the first Indian mathematician to write numbers in the Hindu decimal system with a circle for the zero.

Bhāskara I gave a unique and remarkable rational approximation of the sine function in his commentary on Aryabhata's work. It was written in 629 AD, and it is the oldest known prose work in Sanskrit on mathematics and astronomy.

Bhāskara I and Brahmagupta are two of the most renowned Indian mathematicians who made considerable contributions to the study of fractions.

His work *Mahabhaskariya* contains eight chapters of mathematical astronomy. In chapter 7, he gives a remarkable approximation formula for sin x, which is:

$$\sin x \approx \frac{16x(\pi - x)}{5\pi^2 - 4x(\pi - x)}, \qquad (0 \leq x \leq \frac{\pi}{2})$$

According to Ref 25, "It reveals a relative error of less than 1.9% (the greatest deviation $\frac{16}{5\pi} - 1 \approx 1.859$ at $x = 0$)"

It is not clear, how the above formula relates to the greatest deviation and a relative error at $x = 0$. Therefore, we decided to check the accuracy of this formula for a few values of x:

1. Let's $x = \pi/2 = 90°$

The left-hand part of the formula: $\sin 90° = 1$
The right-hand part of the formula's numerator:

$$16\pi/2 \times (\pi - \pi/2) = 8\pi \times \pi/2 = 4\pi^2$$

The right-hand part of the formula's denominator:

$$5\pi^2 - 4 \times \pi/2 \times (\pi - \pi/2) = 5\pi^2 - 2\pi \times \pi/2 = 5\pi^2 - \pi^2 = 4\pi^2$$

The right-hand part of the formula: $4\pi^2/4\pi^2 = 1$
Thus, $\sin(\pi/2) = \sin 90° = 1$
Conclusion: the accuracy of the formula at $x = \pi/2$ is 100%.

2. Let's $x = \pi/3 = 60°$

The left-hand part of the formula: $\sin 60° \approx \underline{0.8660}$
The right-hand part of the formula's numerator:

$$16\pi/3 \times (\pi - \pi/3) = 16\pi/3 \times 2\pi/3 = 32\pi^2/9$$

The right-hand part of the formula's denominator:

$$5\pi^2 - 4 \times \pi/3 \times (\pi - \pi/3) = 5\pi^2 - 4\pi/3 \times (\pi - \pi/3) =$$
$$5\pi^2 - 4\pi/3 \times 2\pi/3 = 5\pi^2 - 8\pi^2/9 = (9 \times 5\pi^2 - 8\pi^2)/9 =$$
$$(45\pi^2 - 8\pi^2)/9 = 37\pi^2/9$$

The right-hand part of the formula: $(32\pi^2/9)/(37\pi^2/9) = 32/37 \approx \underline{0.8649}$

Thus, sin ($\pi/3$) = sin 60° ≈ <u>0.8660</u>
Conclusion: the accuracy of the formula at $x = \pi/3$ is 0.8649 / 0.8660 = 99.9%.

3. Let $x = \pi/4 = 45°$

The left-hand part of the formula: sin 45° ≈ <u>0.7071</u>
The right-hand part of the formula's numerator:

$$16\pi/4 \times (\pi - \pi/4) = 16\pi/4 \times 3\pi/4 = 48\pi^2/16 = 3\pi^2$$

The right-hand part of the formula's denominator:

$$5\pi^2 - 4 \times \pi/4 \times (\pi - \pi/4) = 5\pi^2 - 4\pi/4 \times (\pi - \pi/4) =$$
$$5\pi^2 - 4\pi/4 \times 3\pi/4 = 5\pi^2 - 12\pi^2/16 = (16 \times 5\pi^2 - 12\pi^2)/16 =$$
$$(80\pi^2 - 12\pi^2)/16 = 68\pi^2/16 = 17\pi^2/4$$

The right-hand part of the formula: $3\pi^2/(17\pi^2/4) = 3 \times 4/17 = 12/17$ ≈ <u>0.7059</u>
Thus, sin ($\pi/4$) = sin 45° ≈ <u>0.7071</u>
Conclusion: the accuracy of the formula at $x = \pi/4$ is 0.7059 / 0.7071 = 99.8%.

4. Let $x = \pi/6 = 30°$

The left-hand part of the formula: sin 30° = <u>0.5</u>
The right-hand part of the formula's numerator:

$$16\pi/6 \times (\pi - \pi/6) = 16\pi/6 \times 5\pi/6 = 80\pi^2/36 = 20\pi^2/9$$

The right-hand part of the formula's denominator:

$$5\pi^2 - 4 \times \pi/6 \times (\pi - \pi/6) = 5\pi^2 - 4\pi/6 \times (\pi - \pi/6) =$$
$$5\pi^2 - 4\pi/6 \times 5\pi/6 = 5\pi^2 - 20\pi^2/36 = (36 \times 5\pi^2 - 20\pi^2)/36 =$$
$$(180\pi^2 - 20\pi^2)/36 = 160\pi^2/36 = 40\pi^2/9$$

The right-hand part of the formula:
$(20\pi^2/9)/(40\pi^2/9) = (20\pi^2 \times 9)/(40\pi^2 \times 9) = \underline{0.5}$
Thus, $\sin(\pi/6) = \sin 30° = \underline{0.5}$
Conclusion: the accuracy of the formula at $x = \pi/6$ is 100%.

Bhāskara I dealt with equations that today are called the Pell's[1] equation.

Bhāskara I posed the problem: "Tell me, a mathematician, what is that square, which multiplied by 8 becomes—together with unity—a square?"

In modern mathematical notation, he asked for the solutions of the equation:

$8x^2 + 1 = y^2$

It has the simple solution: $x = 1$, $y = 3$, or shortly $(x, y) = (1, 3)$ and $(x, y) = (6, 17)$.

Let's solve this equation:

If $x = 1$ and $y = 3$, we have $8 \times 1^2 + 1 = 9 = 3^2$
If $x = 6$ and $y = 17$, we have $8 \times 6^2 + 1 = 8 \times 36 + 1 = 288 + 1 = 289 = 17^2$

---

[1] John Pell (1611–1685) was an English mathematician. He is well known for the Pell's equation $x^2 - ny^2 = 1$, where n is a given positive nonsquare integer (2, 3, 5, 6, 7, 8…, but not 4, 9, 16…, because: $\sqrt{4} = 2$, $\sqrt{9} = 3$, $\sqrt{16} = 4$).

6.5 Mahavira (c. 800–c. 870 AD)

Mahavira (or Mahaviracharya, "Mahavira the Teacher") was a 9th-centuty Jain[2] mathematician (Ref 40).

In his work, astrology was separated from mathematics. It was the earliest Indian text entirely devoted to mathematics. Mahavira explained the same subjects on which Aryabhata and Brahmagupta argued, but he expressed them more clearly. His work is a highly abbreviated approach to algebra and the emphasis in much of his text is on developing the techniques necessary to solve algebraic problems. He is highly respected among Indian mathematicians, because of his formulation of terminology for concepts such as equilateral and isosceles triangle, rhombus, circle, and semicircle. Mahavira's authority spread in all southern India, and his books proved inspirational to other Indian's mathematicians.

Mahavira discovered algebraic identities. As an example, there is one:

$$a^3 = a\,(a + b)\,(a - b) + b^2\,(a - b) + b^3.$$

---

[2] Jain refers to Jainism. Jainism is an ancient Indian religion that prescribes the path of non-violence toward all leaving beings

To prove the above equation, let's to solve the right-hand part of this equation:

$$\text{Right-hand} = a(a+b)(a-b) + b^2(a-b) + b^3 = (a^2+ab)(a-b) + ab^2 - b^3 + b^3 = a^3 + a^2b - a^2b - ab^2 + ab^2 = a^3$$

Proof: the left-hand part of this equation (a^3) is equal to the right-hand part of it (a^3).

Mahavira derived formulas, which approximated area and perimeters of ellipses and found methods to calculate the square of a number and cube roots of a number. He asserted that the square root of a negative number did not exist. He also found out that the sum of the following fractions is approximation of $\sqrt{2}$.

$$1 + \frac{1}{3} + \frac{1}{3\times 4} - \frac{1}{3\times 4\times 34} \approx \sqrt{2}$$

Let's check if this is really the case. To prove this approximation, we do the calculations:

1) $1 + 1/3 + 1/12 = 1 + (4+1)/12 = 1 + 5/12 = (12+5)/12 = 17/12$
2) $17/12 - 1/(3 \times 4 \times 34) = 17/12 - 1/408 = (17 \times 34 - 1)/408$
3) $577/408 = 1.414215686$
$\sqrt{2} = 1.414213562$

Improper fraction 577/408 is greater than $\sqrt{2}$ by only 0.000002124 or 0.00015%.

Mahavira discussed various mathematical topics in his treatise (Ref 41). Among them he described methods to decompose integers and fractions into unit fractions[3]).

Examples are 1/1, 1/2, 1/3, 1/4, 1/5, etc.

Let's do summation of these unit fractions:

$$1/12 + 1/51 + 1/68 = \frac{1\times 17\times 1\times 4\times 1\times 3}{204} = \frac{24}{204} = \frac{2}{17}$$

---

[3] A unit fraction is a rational number written as a fraction, where the numerator is 1, and the denominator is a positive integer. A unit fraction is, therefore, the reciprocal of a positive integer, $1/n$.

As can be seen, the sum of these unit fractions (the right-hand part of the equation) is equal to the left-hand part of the equation.

Mahavira also discussed integer solutions of first-degree indeterminate equation <u>by a method based on the use of the Euclidean algorithm[4]</u>.

Such method of solution occurs in many of the treatises of Indian mathematicians of the classical period, has taken on the more general meaning of "algebra." The following example is an indeterminable linear equation:

> "Three merchants find a purse lying in the road. One merchant says, "If I keep the purse, I shall have twice as much money as the two of you together." "Give me the purse and I shall have three times as much," said the second merchant. The third merchant said, "I shall be much better off than either of you if I keep the purse. I shall have five times as much as the two of you together." How much money does each merchant have?"

If the first merchant has x, the second y, the third z, and p is the amount of money in the purse, then, in modern mathematical notation, the linear equations are:

$$p + x = 2(y + z), \; p + y = 3(x + z), \; p + z = 5(x + y)$$

There is no unique solution. The smallest amount of money in positive integers are:

$x = 1, y = 3, z = 5$, and $p = 15$.

---

[4] In mathematics, the Euclidian algorithm is an efficient method for computing the greatest common divisor (GCD) of two numbers, the largest number that divides both of them without leaving a remainder.

Mahavira formulated rules for calculating combinations and permutations. In modern mathematical notation, such formulas are for calculating combinations (1) and for calculating permutations (2).

$$^nC_r = \frac{n!}{r!(n-r)} \qquad (1)$$

^nC_r is the number of possible combinations of r objects from a set of n objects. The factorial function (symbol "!") means to multiply a series of descending natural numbers.

Example 1. How many combinations of 3 objects (r) from a set of 9 objects (n)?

$$^nC_r = \frac{9\times8\times7\times6\times5\times4\times3\times2\times1}{3!\times6!} = \frac{9\times8\times7\times6\times5\times4\times3\times2\times1}{3!\times6\times5\times4\times3\times2\times1} = \frac{9\times8\times7}{3\times2\times1} = \frac{504}{6} = 84$$

There are 84 combinations.

Formula for the number of possible permutations of k objects from a set of n objects:

$$^nP_k = \frac{n!}{(n-k)!} \qquad (2)$$

Example 2: How many permutations of 3 objects (k) from a set of 9 objects (n)?

$$^nP_k = \frac{n!}{(n-k)!} = \frac{9!}{6!} = \frac{9\times8\times7\times6\times5\times4\times3\times2\times1}{6\times5\times4\times3\times2\times1} = 504$$

There are 504 permutations.

6.6 Bhaskara II (1114–1185)

Bhaskara, (known as "Bhaskara the teacher", and as Bhaskara II to avoid confusion with Bhaskara I), was a mathematician and astronomer (Ref 42).

Bhaskara II and his works represent a significant contribution to mathematical and astronomical knowledge in the 12th century.

Bhaskara II has been called the greatest mathematician of medieval India. His main work, "Crown of Treatises," is divided into four parts, which are considered four independent works. These works deal with arithmetic, algebra, mathematics of the planets, and spheres.

Bhaskara's II work on calculus predates Newton and Leibniz by over 500 years. He is particularly known in the discovery of the principles of differential calculus and its application to astronomical problems and computations. While Newton and Leibniz have been credited with differential and integral calculus, there is strong evidence to suggest that Bhaskara II was in some of the principals of differential calculus. He was the first to conceive the differential coefficient and differential calculus.

Some of Bhaskara's II contributions to mathematics include the following:

- Solutions of quadratic[5], cubic[6], and quartic[7] equations.
- Solutions of indeterminate quadratic equations of the form $ax^2 + b = y^2$.
- Solutions of indeterminate quadratic equations of the form $ax^2 + bx + c = y$ by using a cyclic Chakravala[8] method.
- Method for solutions of equations of the second order, such as $x^2 - ny^2 = 1$.
- Solutions of equations of the second order, such as $61x^2 + 1 = y^2$. This very equation was posed as a problem in 1657 by Pierre de Fermat (1601–1665), a French mathematician, but its solution was unknown in Europe until the time of Leonhard Euler (1717–1783), a Swiss, German, and Russian mathematician.
- Discovered spherical trigonometry that deals with the relationships between the sides and angles of the spherical triangles defined by a number of intersecting great circles on the sphere (a great circle divides the sphere in two equal hemispheres). Spherical trigonometry is of great importance for calculations in astronomy, geodesy and navigation.

Bhaskara II studied the problems leading to more than one solution (Ref 29). Example: *Inside a forest, a number of apes equal to the square of one-eighth of the total apes in the pack are playing noisy games. The remaining twelve apes, who are of a more serious disposition, are on a nearby hill and irritated by the shrieks coming from the forest. What is the total number of apes in the pack?*

---

[5] Quadratic equation is any equation having the form $ax^2 + bx + c = 0$, where b, c are real numbers.

[6] Cubic equation is any equation having the form $ax^3 + bx^2 + cx + d = 0$.

[7] Quartic equation, or equation of the fourth degree, is an equation of the form $a^4 + bx^3 + cx^2 + dx + e = 0$.

[8] Chakravala method is a cyclic algorism to solve indeterminate quadratic equations. (Chakra meaning "wheel" in Sanskrit, the primary language of Hinduism).

According to Bhaskara II, this problem leads to a quadratic equation, and the two solutions, namely 16 and 48, are equally admissible. In modern mathematical notation, this quadratic equation is defined as:

$$ax^2 - bx + c = 0, \text{ where } a = 1/8, b = 1, \text{ and } c = 12$$

Let's solve this equation

$$(\tfrac{1}{8}x)^2 - x + 12 = \tfrac{1}{64}x^2 - x + 12 =$$

$$x^2 - 64x + 12 \times 64 =$$

$$x^2 - 64x + 768 = 0$$

where $a = 1$, $b = 64$, and $c = 768$

We are solving the above equation by using the Brahmagupta formulas for finding the two roots of a quadratic equation (see Section 6.3):

$$x_1 = \frac{64 + \sqrt{64^2 - 4 \times 768}}{2} = \frac{64 + \sqrt{4096 - 3072}}{2} = \frac{64 + \sqrt{1024}}{2} = \frac{64 + 32}{2} = 48$$

$$x_2 = \frac{64 + \sqrt{64^2 - 4 \times 768}}{2} = \frac{64 - \sqrt{4096 - 3072}}{2} = \frac{64 - \sqrt{1024}}{2} = \frac{64 - 32}{2} = 16$$

Our calculations showed that "the two solutions, namely 16 and 48" are correct.

Bhaskara II was more interested in trigonometry for its own sake than his predecessors, who saw it as a tool for calculation (Ref 43).

He obtained many important results among which are several trigonometric formulas. Formulas (1) and (2) are some of them. Currently, they are included in the curriculum of mathematics in the middle and high schools worldwide.

$$\sin(\alpha + \beta) = \sin\alpha \times \cos\beta + \cos\alpha \times \sin\beta \qquad (1)$$
$$\sin(\alpha - \beta) = \sin\alpha \times \cos\beta - \cos\alpha \times \sin\beta \qquad (2)$$

The following exercise proves that these formulas are correct. The calculations are performed, assuming that α = 20° and β = 10°.

7The left-hand part of the equation (1) is:

$$\sin(20° + 10°) = \sin 30° = \underline{0.5}$$

The right-hand part of the equation (1) is:

$$\sin 20° \times \cos 10° + \cos 20° \times \sin 10° =$$
$$0.3420201 \times 0.9848078 + 0.9396926 \times 0.1736482 = 0.3368241 + 0.1631759 = \underline{0.5}$$

The left-hand part of the equation (2) is:

$$\sin(20° - 10°) = \sin 10° = \underline{0.1736482}$$

The right-hand part of the equation (2) is:

$$\sin 20° \times \cos 10° - \cos 20° \times \sin 10° =$$
$$0.3420201 \times 0.9848078 - 0.9396926 \times 0.1736482 = 0.3368241 - 0.1631759 = \underline{0.1736482}$$

In many ways Bhaskara II represents the peak of mathematical knowledge in the 12th century. He reached an understanding of the number systems and solving equations, which were not to be achieved in Europe for several centuries. He rightly achieved an outstanding reputation for his remarkable contribution.

6.7 Madhava of Sangamagrama (1340–1425)

Madhava of Sangamagrama was a mathematician and astronomer from the town of Sangamagrama in South India. He is considered the founder of the Kerala School of astronomy and mathematics (Ref 44).

One of the greatest mathematician-astronomers of the medieval period, Madhava made pioneering contributions to the study of infinite series, calculus, trigonometry, geometry, and algebra.

Some scholars have suggested that Madhava's work may have been transmitted to Europe via Jesuit[9] missionaries and traders who were active around the port of Muziris (modern day Indian state of Kerala) at the time. As a result, it may have had <u>an influence on later European developments in analysis and calculus.</u>

Madhava's contributions to mathematics including infinite series, trigonometry, value of π, algebra, and calculus.

Madhava discovered the infinite series for the trigonometric functions of sine, cosine, and tangent. The text describes the series in the following manner:

"The first term is the product of the given sine and radius of the desired arc divided by cosine of the arc. The succeeding terms are obtained by a process of iteration when the first term is repeatedly multiplied by the square of the sine and divided by the square of the cosine. All the terms are then divided by the odd numbers 1, 3, 5… The arc is

---

[9] The Society of Jesus is a male religious congregation of the Catholic Church. The members are called Jesuits.

obtained by adding and subtracting respectively the terms of odd rank and those of even rank. It is laid down that the sine of the arc or that of its complement whichever is the smaller should be taken here as given sine. Otherwise the terms obtained by this above iteration will not tend to the vanishing magnitude."

This series was traditionally known as the Gregory[10] series. Today, it is referred to as the Madhava – Gregory – Leibniz[11]) series.

Madhava composed an accurate table of sines (the table lists the trigonometric sines of the twenty-four angles 3.75°, 7.50°, 11.25°…, and 90.00°). Making a quarter circle at twenty-four equal interval, he gave the length of the half-chord (sines) corresponding to each of them.

No work of Madhava detailing the methods, which he used for the computation of the sine table, has survived. However from the writing of later mathematicians like Nilakantha Somayaji (1444–1544) and Jyeshtadeva (1500–1575) that give ample references to Madhava's accomplishments. It is conjectured that Madhava computed his sine table using the power series expansion of sin x.

In modern mathematical notation, this power series is defined as:

$$\sin x = x - \frac{x^3}{3!} + \frac{x^5}{5!} - \frac{x^7}{7!} + \cdots$$

where x is expressed in degrees.

Each of these power series represent arithmetic fractions. Let's solve this equation with three power series, where $x = 3.75$:

sin 3.75 = 3.75 − 3.75³/3! + 3.75⁵/5! − 3.75⁷/7! =
3.75 − 52.73/6 + 741.58/120 − 10428.43/5040 =
3.75 − 8.79 + 6.18 − 2.07 = −093

---

[10] James Gregory (1638–1675), was a Scottish mathematician and astronomer.
[11] Gottfried Wilhelm Leibniz (1646 –1716) was a German polymath and philosopher.

Let's see what happens if we add the fourth power series, in which the numerator is 3.75 to the 9th power, and the denominator is factorial 9 (9!).

$$3.75^9 = 146649.78$$
$$9! = 362880$$
$$146649.78 / 362880 = 0.40$$

Let's add 0.40 and solve the hand-right part of this equation:

$$\sin 3.75 = 3.75 - 8.79 + 6.18 - 2.07 + 0.40 = -\underline{0.53}$$

Apparently, the process of adding and subtracting such fractions will take a long time to solve such an equation. At this stage, sin 3.75 = − 0.53.

According to the modern trigonometric table, sin 3.75 = 0.0654. At this stage, we need to find a number (x) that is even to 0.0654.

$$x = 0.0654 + 0.53 = 0.5954$$

So, sin 3.75 = − 0.53 + 0.5954 = 0.0654

Madhava's and modern values of sin x are shown in the following table:

Angle x in degrees	Value of sin x derived from Madhava's table	Modern value of sin x
03.75	0.06540314	0.06540313
07.50	0.13052623	0.13052619
11.25	0.19509032	0.19509032
15.00	0.25881900	0.25881905
18.75	0.32143947	0.32143947
22.50	0.38268340	0.38268343
26.25	0.44228865	0.44228869

30.00	0.49999998	0.50000000
33.75	0.55557022	0.55557023
37.50	0.608761<u>39</u>	0.608761<u>43</u>
41.25	0.65934580	0.65934582
45.00	0.707106<u>81</u>	0.707106<u>78</u>
48.75	0.75183985	0.75183981
52.50	0.79335331	0.79335334
56.25	0.83146960	0.83146961
60.00	0.86602543	0.86602540
63.75	0.89687275	0.89687274
67.50	0.92387954	0.92387953
71.25	0.94693016	0.94693013
75.00	0.96592581	0.96592583
78.75	0.98078527	0.98078528
82.50	0.99144487	0.99144486
86.25	0.99785895	0.99785892
90.00	0.99999997	1.00000000

Comparison between the values of sin x derived from Madhava's table and that of modern calculations, shows that there is no difference in the first seven decimal places for 19 various angles, and no difference in the first six decimal places only for 3 angles (the last two decimal places are underlined). The accuracy of calculated values of sine are very impressive.

Madhava's work on the value of π is sited in the "Methods for the great sines."

$$\frac{\pi}{4} = 1 - \frac{1}{3} + \frac{1}{5} - \frac{1}{7} + \cdots + \frac{(-1)^n}{2n+1} + \cdots$$

It gives the infinite series expansion of π, now known as the Madhava–Leibnitz series.

Madhava gave a correction term (R_n) for the error after computing the sum up to *n* terms.

$$R_n = (n^2 + 1)/(4n^3 + 5n)$$

Increase of the *n* terms, increases the accuracy in calculating of π.
Madhava also gave a more rapidly converging series by transforming the original infinite series of π.

$$\pi = \sqrt{12}\left(1 - \frac{1}{3\cdot 3} + \frac{1}{5\cdot 3^2} - \frac{1}{7\cdot 3^3} + \cdots\right)$$

Let's calculate the value of π from the above equation in five steps:

1) $\sqrt{12} = 3.4641016$
2) $1 - 1/9 = 8/9$
3) $8/9 + 1/45 = 41/45$
4) $41/45 - 1/189 = 7704/8505 = 0.9058201$
5) $3.4641016 \times 0.9058201 = \underline{3.1378528}$.

π = 3.1415926 (with seven decimal places) is greater than the value of π obtained from the above original infinite series for only:

$$3.141592654 - 3.1378528 = 0.003739854$$

This means that π is large only by 0.119%.

By using the first 21 terms to compute an approximation of π, Madhava obtained a value correct to 11 decimal places: 3.14159265359. The value of π correct to 13 decimal places is also attributed to Madhava. Mathematicians reached an accuracy of 35 decimal places by the beginning of the 17th century. Madhava laid the foundations for the development some components of calculus such as differentiation, term-by-term integration, iterative methods for solutions of nonlinear equations, and the theory that the area under a curve is its integral.

6.8 Nilakantha Somayaji (1444–1544)

Nilakantha Somayaji, known as Kelallur Comatiri, was a major mathematician and astronomer of the Kerala School of astronomy and mathematics (Ref 45).

One of his most influential work was the comprehensive astronomical treatise *Tantrasamgraha*. He also composed an elaborate commentary called the *Aryabhatiya Bhasya*, in which he had discussed infinite series expansions of trigonometric functions and problems of algebra and spherical geometry.

The *Tantrasamgraha* is very important in terms of the mathematics Nilakantha used (Ref 46). In particular he used results discovered by Madhava and it is an important source of the remarkable mathematical results which he discovered. However, Nilakantha did not just use Madhava's results, he extended them and improved them.

The series π/4 = 1 − 1/3 + 1/5 − 1/7 + ... is a special case of the series representation for arctangent[12]): tan−1(x) = x − x3/3 + x5/5 − x7/7

The left-hand part of the series is: π/4 = 0.785398.
The right-hand part has the series of eight members is:

1 − 1/3 + 1/5 − 1/7 + 1/9 − 1/11 + 1/13 − 1/15 + ... =
1 − 0.333333 + 0.2 − 0.142857 + 0.111111 − 0.090909 + 0.076923 − 0.066667

Let's summarize this eight members of the series:

1) 1 − 0.333333 = 0.666667
2) 0.666667 + 0.2 = 0.866667
3) 0.866667 − 0.142857 = 0.723810
4) 0.723810 + 0.111111 = 0.834921
5) 0.834921 − 0.090909 = 0.744012
6) 0.744012 + 0.076923 = 0.820935
7) 0.820935 − 0.066667 = 0.754268

The sum of numbers in the right-hand part of the series (0.754268) less than the number on the left-hand part of the series (0.785398):

0.785398 − 0.754268 = 0.031130 (the difference is 4%)

Let's add the following series of the four numbers:

1/17 − 1/19 + 1/21 − 1/23 = 0.058824 − 0.052632 + 0.047619 − 0.043478

---

[12] Arctangent is one of the inverse trigonometric functions: sine, cosine, tangent, cotangent, secant, and cosecant functions. Inverse trigonometric functions are widely used in engineering, navigation, physics, and geometry.

Let's summarize this four members of the series:

1) 0.754268 + 0.058824 = 0.813092
2) 0.813092 − 0.052632 = 0.760460
3) 0.760460 + 0.047619 = 0.808079
4) 0.808079 − 0.043478 = <u>0.764601</u>

The sum of numbers in the right-hand part of the series (0.764601) less than the number on the left-hand part of the series (0.785398):

0.785398 − 0.764601 = <u>0.020797</u> (the difference is 3%)

If you want to continue this process, add the series of the following four numbers:

1/25 − 1/27 + 1/29 − 1/31 = 0.04 − 0.037037 + 0.034483 − 0.032258

6.9 Jyesthadeva (1500–1575)

Jyesthadeva was an astronomer-mathematician of the Kerala School of astronomy and mathematics (Ref 47). He is best known as the author of *Yuktibhasa*, a survey of Kerala mathematics. His work contains proofs of the theorems and gives derivations of the rules. It was very unusual for an Indian mathematical text at the time. This is one of the main mathematical and astronomical texts produced by the Kerala School.

The *Yuktibhasa* is very important in terms of the mathematics presented by Jyesthadeva. In particular, he presents the remarkable mathematical theorems discovered by Madhava. Written in about 1550, Jyesthadeva's commentary contained proofs of the earlier results by Madhava and Nilakantha, who did not give.

The planetary theory presented by Jyesthadeva is similar to that adopted by Tycho Brahe[13] a century later.

Three factors make Jyesthadeva's book *Yuktibhasa* unique in the history of the development of mathematical thinking in the Indian subcontinent:

- It is composed in the spoken language of the local people, namely, the Malayalam language. This is in contrast to the centuries old Indian tradition of composing scholarly works in the Sanskrit language, which was the language of the learned.
- The work is in prose, in contrast to the prevailing style of writing even technical manuals in verse. All the other notable works of the Kerala School are in verse.
- Most importantly, *Yuktibhasa* was composed intentionally as a manual of proofs. The very purpose of writing the book was to record in full detail the rationales of the various results discovered by mathematicians–astronomers of the Kerala School, especially of Nilakantha Somayaji. This book is proof enough to establish that the concept of proof was not unknown to Indian mathematical traditions.

---

[13] Tycho Brahe, born Tyge Ottesen Brahe (1546–1601), was a Danish nobleman known for his accurate and comprehensive astronomical and planetary observations.

CHAPTER 7

ISLAMIC MATHEMATICS

The Islamic Empire (Ref 2) established across Persia, the Middle East, Central Asia, North Africa, Iberia, and in parts of India in the 8th century made significant contribution towards mathematics. Islamic texts on mathematics were written in Arabic, but most of them were not written by Arabs. Arabic was used as the written language of non-Arabic scholars throughout the Islamic world at the time. Persians contributed to the world of mathematics alongside Arabs.

Mathematics during the Golden Age[1] of Islam, especially during the 9th and 10th centuries, was built on Greek mathematics and Indian mathematics.

Important progress (Ref 12) was made, such as the full development of the decimal place-value system to include decimal fractions, the first systematized study of algebra (named for *The Compendious Book on Calculation by Completion and Balancing* by scholar Al-Khwarizmi), and advances in geometry and trigonometry. Arabic works also played an important role in the transmission of mathematics to Europe during the 10th to 12th centuries.

One consequence of the Islamic prohibition on depicting the human form was the extensive use of complex geometric patterns to decorate their buildings, raising mathematics to the form of an art.

---

[1] The Islamic Golden Age is the era, traditionally dated from the 8th century to the 14th century.

In fact, over time, Muslim artists discovered all the different forms of symmetry that can be depicted on a two-dimensional surface.

The Quran itself encouraged the accumulation of knowledge, and a Golden Age of Islamic science and mathematics flourished throughout the medieval period from the 9th to 15th Centuries. The House of Wisdom was set up in Baghdad around 810, and work started almost immediately on translating the major Greek and Indian mathematical and astronomy works into Arabic.

With the stifling influence of the Turkish Ottoman Empire from the 14th or 15th Century onwards, Islamic mathematics stagnated, and further developments moved to Europe.

7.1 Muhammad ibn Musa al-Khwarizmi (c. 780–c. 850)

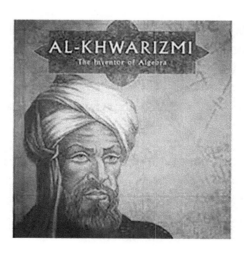

Muhammad ibn Musa al-Khwarizmi (Ref 48) was a Persian mathematician, astronomer, geographer, and a scholar in the House of Wisdom in Baghdad during the Abbasid Caliphate. In the 12th century, Latin translation of his work on the Indian numerals introduced the decimal positional number system to the Western world.

Al-Khwarizmi's *The Compendious Book on Calculation by Completion and Balancing* presented the first systematic solution of linear and quadratic equations in Arabic. The book was a source of information for spreading the Hindu-Arabic numeral system throughout the Middle East and Europe. It was translated into Latin as *Algoritmi de numero Indorum*. Al-Khwarizmi, rendered as *Algoritmi*, the Latin form of his name.

Al-Khwarizmi is often considered one of the fathers of algebra. The term "algebra" is derived from *al-jabr*, one of the two operations he used to solve quadratic equations (*al-jabr*, meaning "restoration", referring to adding a number to both sides of the equation to consolidate or cancel terms).

Al-Khwarizmi's method of solving linear and quadratic equations worked by first reducing the equation to one of six standard forms (b and c are positive integers):

- Squares equal roots ($ax^2 = bx$)
- Squares equal number ($ax^2 = c$)
- Roots equal number ($bx = c$)
- Squares and roots equal number ($ax^2 + bx = c$)
- Squares and number equal roots ($ax^2 + c = bx$)
- Roots and number equal squares ($bx + c = ax^2$)

In al-Khwarizmi's day, these notations had not yet been invented, so he had to use ordinary text to present problems and their solutions. For example, there is one problem he described in his book.

If someone says, "You divide ten into two parts: multiply the one by itself; it will be equal to the other taken eighty-one times." Computation: You say, ten less thing, multiplied by itself, is a hundred plus a square less twenty things, and this is equal to eighty-one things. Separate the twenty things from a hundred and a square and add them to eighty-one. It will then be a hundred plus a square, which is equal to a hundred and one roots. Halve the roots; the moiety is fifty and a half. Multiply this by itself, it is two thousand five hundred and fifty and a quarter. Extract the roots from this; it is forty-nine and a half. Subtract this from the moiety of roots, which is fifty and a half. There remains one, and this is one of two parts.

In modern mathematical notation, this process with x the *thing* or *root*, is given as $(10 - x)^2 = 81x$.

Let's solve this equation

$(10 - x)^2 - 81x = 0$
$(100 - 2 \times 10 \times x + x^2) - 81x = 0$
$100 - 20x + x^2 - 81x = 0$ and finally:
$x^2 - 101x + 100 = 0$, which is a quadratic equation.

The roots of this quadratic equation are calculated by the formula:

$$x_{1,2} = \frac{-b \pm \sqrt{b^2 - 4ac}}{2a}$$

where $b = -101$, $a = 1$, $c = 100$

$$x_1 = \frac{101 + \sqrt{101^2 - 4 \times 100}}{2} = \frac{101 + \sqrt{10201 - 400}}{2} = \frac{101 + 99}{2} = 100$$

$$x_2 = \frac{101 - \sqrt{101^2 - 4 \times 100}}{2} = \frac{101 - \sqrt{10201 - 400}}{2} = \frac{101 - 99}{2} = 1$$

This quadratic equation has two roots: $x_1 = 1$ and $x_2 = 100$. The problem described and solved by al-Khwarizmi has only one root $x = 1$.

J. J. O'Conner and E. F. Robertson wrote in the *MacTutor History of Mathematics archive*: "Perhaps one of the most significant advances made by Arabic mathematics began at this time with the work of al-Khwarizmi, namely the beginnings of algebra. It is important to understand just how significant this new idea was. It was a revolutionary move away from the Greek concept of mathematics, which was essentially geometry. Algebra was a unifying theory, which allowed rational numbers, irrational numbers, geometrical magnitudes, etc., to all be treated as "algebraic objects." It gave mathematics a whole new development path so much broader in concept to that, which had existed before, and provided a vehicle for future development of the subject. Another important aspect of the introduction of algebraic ideas was that it allowed mathematics to be applied to itself in a way, which had not happened before."

Some of al-Khwarizmi's work was based on Persian and Babylonian astronomy, Indian numbers, and Greek mathematics.

When, in the 12th century, his works spread to Europe through Latin translations, it had a profound impact on the advance of mathematics in Europe.

The word *algorithm* (Ref 2) is derived from the Latinization of the name al-Khwarizmi, Algoritmi.

In mathematics and computer science, algorithm is a self-contained step-by-step set of operations to be performed. Algorithms perform calculations, data processing, and automated reasoning tasks.

7.2 Abu Kamil Shuja ibn Aslam (c. 850–c. 930)

Abu Kamil Shuja ibn Aslam was an Egyptian Muslim mathematician during the Islamic Golden Age (Ref 49). He is considered the first mathematician to systematically use and accept irrational numbers as solutions and coefficients to equations.

Almost nothing is known about the life and career of Abu Kamil except that he was a successor of al-Khwarizmi, whom he never personally met.

Abu Kamil made important contributions to algebra and geometry. He was the first Islamic mathematician to work easily with algebraic equations with powers higher than x^2 (up to x^8), and solved sets of non-linear simultaneous equations[2] with three unknown variables.

Abu Kamil worked with powers of the unknown higher than x^2. These powers were not given in symbols but were written in words (Ref 50). For example, he used the expressions: "square-square-root," "cube-cube," and "square-square-square-square," which, in modern mathematical notation, are:

$$x^2 \times x^2 \times x = x^5$$
$$x^3 \times x^3 = x^6$$
$$x^2 \times x^2 \times x^2 \times x^2 = x^8$$
$$x^n \times x^m = x^{n+m}$$

Abu Kamil extended algebra to the set of irrational numbers, accepting square roots and fourth roots as solutions and coefficients to quadratic equations. His works formed an important foundation for the development of algebra and influenced later mathematicians, such as al-Karaji and Fibonacci (Ref 2).

---

[2] In mathematics, a set of simultaneous equations, also known as a system of equations, is a finite set of equations for which common solutions are sought.

7.3 Al-Karaji (953–1029)

Abu Bakr ibn Muhammad ibn al Husayn al-Karaji was Iranian mathematician and engineer (Ref 51). His three surviving principal works are mathematical: *Al-Badi' fi'l-hisab* (*Wonderful on calculation*), *Al-Fakhri fil-jabr wa'l-muqabala* (*Glorious on algebra*), and *Al-Kafi fi'l-hisab* (*Sufficient on calculation*).

Some historians consider Al-Karaji to be merely reworking the ideas of other (he was influenced by Diophantus), but most regard him as original, in particular for the beginnings of freeing algebra from geometry.

He studied the algebra of exponents, and was the first to realize that the sequence x, x^2, x^3... could be extended indefinitely, as well as the reciprocals $1/x$, $1/x^2$, $1/x^3$... His work on algebra and polynomials gave the rules for arithmetic operations: adding, subtracting, and multiplying polynomials; but was restricted to dividing polynomials by monomials.

Al-Karaji also uses a form of mathematical induction in his arguments, although he certainly does not give a rigorous exposition of the principle (Ref 52).

Basically what al-Karaji does is to demonstrate an argument for $n = 1$, then prove the case $n = 2$ based on his result for $n = 1$, then prove the case $n = 3$ based on his result for $n = 2$, carry on to around $n = 5$ before remarking that one can continue the process indefinitely. Although this is not induction proper, it is a major step towards understanding inductive proofs. Al-Karaji in *Glorious on algebra* computed $(a + b)^3$ in *Wonderful on calculation* he computed $(a - b)^3$ and $(a + b)^4$. He constructed a table in which he presented a tabular arrangement of the binomial coefficients:

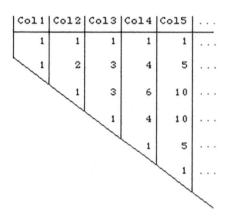

Al-Karaji wrote, "In order to succeed we must place *one* on a table and *one* below the first *one*, move the first *one* into a second column, add the first *one* to the *one* below it. Thus we obtain *two*, which we put below the transferred *one*, and we place the second *one* below the *two*. We have therefore *one*, *two*, and *one*."

To see how the second column of 1, 2, 1 corresponds to squaring $(a + b)$, al-Karaji wrote, "This shows that for every number composed of two numbers, if we multiple each of them by itself once – since the two extremes are *one* and *one* – and if we multiply each one by the other twice – since the intermediate term is *two* – we obtain the square of this number."

In modern mathematical notation, this statement can be expressed as a binomial $(a + b)^2$. Its expansion is:

$$(a + b)^2 = (a + b)(a + b) = a^2 + ab + ba + b^2 = a^2 + 2ab + b^2$$

Al-Karaji described the third column of 1, 3, 3, 1: "If we transfer the *one* in the second column into a third column, then add *one* from the second column to *two* below it, we obtain *three* to be written under the *one* in the third column. If we then add *two* from the second column to *one* below, we have *three* which is written under the *three*, then we write *one* under this *three*; we thus obtain a third column whose numbers are *one*, *three*, *three*, and *one*."

In modern mathematical notation, this statement can be expressed as a binomial $(a + b)^3$. Its expansion is:

$$(a + b)^3 = (a + b)^2 (a + b) = (a^2 + 2ab + b^2)(a + b) =$$
$$a^3 + \underline{2a2b} + \underline{ab2} + \underline{a2b} + \underline{2ab2} + b^3 = a^3 + 3a^2b + 3ab^2 + b^3$$

Description of the table continues up to column 5 where binomial coefficients are obtained by expansion of $(a + b)^5$.

$$(a + b)^5 = (a + b)^3 (a + b)^2 = (a^3 + 3a^2b + 3ab^2 + b^3)(a^2 + 2ab + b^2) =$$
$$a^5 + \underline{3a4b} + \underline{3a3b2} + \underline{a2b3} + \underline{2a4b} + \underline{6a3b2} + \underline{6a2b3} + \underline{2ab4} + \underline{a3b2}$$
$$+ \underline{3a2b3} + \underline{3ab4} + b^5 = a^5 + 5a^4b + 10a^3b^2 + 10a^2b^3 + 5ab^4 + b^5$$
$$(a + b)^5 = a^5 + 5a^4b + 10a^3b^2 + 10a^2b^3 + 5ab^4 + b^5$$

Al-Karaji developed a method for summation of the first n natural numbers, such as $1 + 2 + 3 + \ldots + n$. In modern mathematical notation, his method is expressed by the formula:

$$\Sigma n = \frac{n(n+1)}{2} \tag{1}$$

Let's check if the formula (1) is correct. As an example, the numbers from 1 to 5 are taken in natural order.

The left-hand part of the formula gives:

$$1 + 2 + 3 + 4 + 5 + 6 + 7 + 8 + 9 + 10 = 55$$

The right-hand part of the formula gives:

$$10 \times (10 + 1)/2 = 10 \times 11/2 = 110/2 = 55$$

The formula (1) is correct.

Al-Karaji worked on summation of the square of the number. He wrote, "The sum of the squares of the numbers that follow one another in natural order from one is equal to the sum of these numbers and the product of each of them by its predecessor." In modern mathematical notation this result is expressed by the formula:

$$\Sigma n^2 = \Sigma n + \Sigma n (n - 1) \tag{2}$$

Let's check if the formula (2) is correct. As an example, the numbers from 1 to 5 are taken in natural order.

The left-hand part of the formula gives:

$$(1 + 2 + 3 + 4 + 5)^2 = 15^2 = \underline{225}$$

The right-hand part of the formula gives:

$$(1 + 2 + 3 + 4 + 5) + (1 + 2 + 3 + 4 + 5) \times [(1 + 2 + 3 + 4 + 5) - 1] = 15 + 15 \times (15 - 1) = 15 + 15 \times 14 = 15 + 210 = \underline{225}$$

The formula (2) is correct.

Al-Karaji also considered sums of the cubes of the first n natural numbers. He wrote, "If we want to add the cubes of the numbers that follow one another in their natural order we multiply their sum by itself." In modern mathematical notation, this statement can be expressed by the formula:

$$\Sigma n^3 = (\Sigma n)^2 \tag{3}$$

Let's check if the formula (3) is correct. As an example, the numbers from 1 to 5 are taken in natural order.

The left-hand part of the formula gives:

$$1^3 + 2^3 + 3^3 + 4^3 + 5^3 = 1 + 8 + 27 + 64 + 125 = \underline{225}$$

The right-hand part of the formula gives:

$(1 + 2 + 3 + 4 + 5)^2 = 15^2 = \underline{225}$

The formula (3) is correct.

Besides algebra, al-Karaji discussed some of his geometrical work. He defined points, lines, angles, surfaces, and solids. He also developed methods of measuring plane and solid figures.

7.4 Ibn al-Haytham (c. 965–c. 1040)

Abu Ali al-Hasan ibn al-Haytham, known as Alhazen or Alhacen, was an Arab scientist, mathematician, astronomer, and philosopher (Ref 53). Ibn al-Haytham is widely considered to be one of the first theoretical physicists, and an early proponent of the concept that a hypothesis must be proved by experiments or mathematical evidence.

In medieval Europe[3], Ibn al-Haytham (Alhazen) was honored as *Ptolemaeus Secundus* (the "Second Ptolemy") or called "The Physicist". He is also sometimes <u>called al-Basri after his birthplace Basra in Iraq, or al-Misri ("of Egypt")</u>

---

[3] In the history of Europe, the medieval period lasted from the 5th to the 15th century.

In geometry, Alhazen attempted to solve the problem of squaring the circle using the area of lunes (crescent shapes), but later gave up on the impossible task.

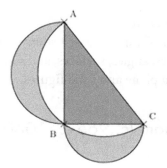

The two lunes formed from a right triangle by erecting a semicircle on each of the triangle's sides, inward for the hypotenuse and outward for the other two sides, are known as the lunes of Alhazen; they have the same total area as the triangle itself.

Alhazen's contributions to number theory include his work on **perfect numbers**. In his *Analysis and Synthesis* work, he may have been the first to state that every **perfect number** can be found by using his method.

Discoverers of the perfect numbers

The 1rst perfect number is **6**. It named perfect because 1, 2, and 3 are its proper positive divisors: 1 + 2 + 3 = 6. The number 6 is equal to half the sum of all its positive divisors:

$$(1 + 2 + 3 + 6)/2 = 6.$$

The 2nd perfect number is **28**:

$$28 = 1 + 2 + 4 + 7 + 14.$$

The 3rd perfect number is **496**:

$$496 = 1 + 2 + 4 + 8 + 16 + 31 + 62 + 124 + 248$$

The 4th perfect numbers is **8128**

These 4 numbers discovered a Greek mathematician Euclid (330–260 BC).

Perfect numbers were preceded by prime numbers. French mathematician Marin Mersenne (1588–1648) was best known by his formula to generate prime numbers. Often prime numbers are named after him, "Mersenne numbers," which calculated by the formula:

$$M_n = 2^n - 1$$

where n is a prime number.

Mersenne believed that his formula was universal, but sometimes it generates not prime numbers, but composite numbers looking like prime numbers. For example:

$$M_n = 2^{11} - 1 = 2048 - 1 = 2047 \text{ and } 2047 = 23 \times 89$$
$$M_n = 2^{23} - 1 = 8{,}388{,}608 - 1 = 8{,}388{,}607 \text{ and } 8{,}388{,}607 = 47 \times 178{,}481$$

Prime and perfect numbers seem to depend on each other. Leonhard Euler proved that all perfect numbers can be calculated by such formula:

$$P = 2^{n-1} \times (2^n - 1)$$

where n and $(2^n - 1)$ are prime numbers.

Let's check if the following formula can generate perfect numbers. The 1st perfect number (P_2)

$$P_2 = 2^{2-1} \times (2^2 - 1) = 2 \times (4 - 1) = 6$$

The 2nd perfect number (P_3)

$$P_3 = 2^{3-1} \times (2^3 - 1) = 4 \times (8 - 1) = 28$$

The 3rd perfect number (P_5)

$$P_5 = 2^{5-1} \times (2^5 - 1) = 16 \times (32 - 1) = 16 \times 31 = 496$$

The 4th perfect number (P_7)

$$P_7 = 2^{7-1} \times (2^7 - 1) = 64 \times (128 - 1) = 64 \times 127 = 8{,}128$$

Let's check if the next number (P_{11}) is a perfect.

$$P_{11} = 2^{11-1} \times (2^{11} - 1) = 1{,}024 \times (2{,}048 - 1) = 1{,}024 \times 2{,}047 = 2{,}096{,}128$$

The 2nd factor of this formula ($2^{11} - 1$) = 2,047. This number is not a prime. It consist of two numbers: 23 and 89 (23 × 89 = 2047). Therefore P_{11} is not a perfect number.

Let's check if the 5th number (P_{13}) is a perfect:

$$P_{13} = 2^{13-1} \times (2^{13} - 1) = 4{,}096 \times (8{,}192 - 1) = 4{,}096 \times 8{,}191 = 33{,}550{,}336$$

The 2nd factor of this formula ($2^{13} - 1$) = 8191. This number is a prime. Therefore P_{13} is a perfect number.

The 5th perfect number 33,550,336 was discovered in 1456. Discoverer of this perfect number is not specified.

Let's check if the 6th number (P_{17}) is a perfect:

$$P_{17} = 2^{17-1} \times (2^{17} - 1) = 65{,}536 \times (131{,}072 - 1) = 65{,}536 \times 131{,}071 = 8{,}589{,}869{,}056$$

The 2nd factor of this formula ($2^{17} - 1$) = 131,071. This number is a prime. Therefore P_{17} is a perfect number.

The 6th perfect number 8,589,869,056 was discovered in 1588 by Italian mathematician Pietro Cataldi (1548–1626).

We used the *TI-30X SOLAR* TEXAS INSTRUMENTS calculator. It allowed to generate perfect numbers that have no more than 10 digits.

To generate perfect numbers, having more than 10 digits, we used the TI-30Xa TEXAS INSTRUMENTS calculator.

Let's check if the 7th number (P_{19}) is a perfect:

$$P_{19} = 2^{19-1} \times (2^{19} - 1) = 2^{18} \times (2^{19} - 1) = 262{,}144 \times (524{,}288 - 1) = 262{,}144 \times 524{,}287 = 137{,}438{,}691{,}328$$

The 2nd factor of this formula $(2^{19} - 1) = 524{,}287$. This number is a prime. Therefore P19 is a perfect number.

The 7th perfect number 137,438,691,328 was discovered in 1588 by Pietro Cataldi.

Here is our list of perfect numbers:

$P_2 = 6$
$P_3 = 28$
$P_5 = 496$
$P_7 = 8{,}128$
$P_{13} = 33{,}550{,}336$
$P_{17} = 8{,}589{,}869{,}056$
$P_{19} = 137{,}438{,}691{,}328$

Getting the next perfect 23-digit number (P_{23}) was not possible using a regular home computer.

We were interested why there is such a big gap between the perfect number P_{19} and the perfect number P_{31}. Probably, because the numbers P_{23} and P_{29} are not perfect.

Let's check:

$$P_{23} = 2^{23-1} \times (2^{23} - 1) = 4{,}194{,}304 \times (8{,}388{,}608 - 1) = 4{,}194{,}304 \times \underline{8{,}388{,}607}$$

If the second factor of this formula (underlined) is a prime number, then P23 is a perfect number. It turns out that it is composite: 8,388,607 = 47 × 178,481. Therefore P_{23} is not a perfect number.

Let's repeat the same with the number P29:

$$P_{29} = 2^{29-1} \times (2^{29} - 1) = 268{,}435{,}456 \times (536{,}870{,}912 - 1) = 268{,}435{,}456 \times \underline{536{,}870{,}911}$$

If the second factor of this formula (underlined) is a prime number, then P_{29} is perfect number. It turns out that it is composite: 536,870,911 = 233 × 1103 × 2089. Therefore the number P_{29} is not a perfect number.

The 8th perfect number is P_{31} (19 digits) was discovered by Leonhard Euler in 1772. Detailed information about Euler is given in Chapter 10.7.

The 9th perfect number is P_{61} (37 digits) was discovered in 1883 by Ivan Mikheevich Pervushin (1827–1900) a Russian clergyman and a mathematician.

The 10th and 11th perfect numbers are P_{89} (54 digits) and P_{107} (65 digits) were discovered in 1911 and 1914 by Ralf Ernest Powers (1875–1952). He was an American amateur mathematician.

Currently, we have the 51st perfect number discovered in December 2018. This number has 49,724,095 digits.

7.5 Omar Khayyam (May 18, 1048–December 4, 1131)

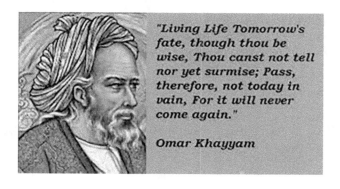

"*Living Life Tomorrow's fate, though thou be wise, Thou canst not tell nor yet surmise; Pass, therefore, not today in vain, For it will never come again.*"

Omar Khayyam

Ghiyath ad-Din Abu'l-Fath Umar ibn Ibrahim al-Khayyam Nishapuri, known as Omar Khayyam, was a Tajik Persian mathematician, astronomer, philosopher, and poet, who is widely considered to be one of the most influential scientists of the Middle Ages (Ref 54). He wrote numerous treatises on mechanics, geography, mineralogy and astronomy.

Khayyam was famous during his times as a mathematician. He wrote the influential *Treatise on Demonstration of Problems of Algebra*, which laid down the principles of algebra, part of the body of mathematics that was eventually transmitted to Europe. In particular, he derived general methods for solving cubic equations and even some higher orders. In the *Treatise*, he wrote on the triangular array of binomial coefficients (in modern mathematics, it is known as Pascal's triangle).

In 1077, Khayyam wrote "Explanations of the Difficulties in the Postulates of Euclid" published in English as "On the Difficulties of Euclid's Definitions." Khayyam's attempt was a distinct advance, and his criticisms made their way to Europe, and may have contributed to the eventual development of non-Euclidean geometry.

Omar Khayyam created important works on geometry, specifically on the theory of proportions. In geometric algebra he developed a method of solving cubic equations. His solutions are not numbers but rather line segments[4]. In this regard Khayyam's work can be considered the first systematic study and the first exact <u>method of solving cubic equations.</u>

---

[4] A line segment is a part of a line that is bounded by two distinct end points, and contains every point on the line between its endpoints.

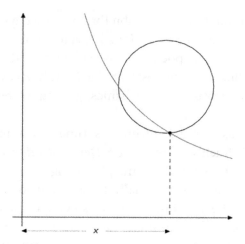

Omar Khayyam's geometric solution to the cubic equation:

$$x^3 + 200x = 20x^2 + 2000$$

Omar Khayyam found a positive root for this equation by intersecting a hyperbola with a circle. This particular geometric solution of cubic equations has been further investigated and extended to degree four equations.

As a mathematician, Khayyam has made fundamental contributions to the philosophy of mathematics especially in the context of Persian Mathematics and Persian philosophy with the other Persian scientists and philosophers.

7.6 Al-Samawal al-Maghribi (c. 1130–c. 1180)

Al-Samawal ibn Yahya al-Maghribi, commonly known as Samau'al al-Maghribi, was a mathematician, astronomer, and physician. Though born to a Jewish family, he concealed his conversion to Islam for many years in fear of offending his father, then openly embraced Islam in 1163 after he had a dream telling him to do so. His father was a Rabbi from Morocco (Ref 55).

Al-Samawal was about 13 years old, when he began serious study, beginning with Hindu methods of calculation and a study of astronomical tables (Ref 56).

Al-Samawal had soon mastered all the mathematics, which his teachers knew. These teachers had covered topics including an introduction to surveying, elementary algebra, and the geometry of the first few books of Euclid's *Elements*.

In order to take his mathematical studies further, al-Samawal had to study on his own. He read the works of Abu Kamil, al-Karaji and others so that by the time he was eighteen years old he had read almost all the available mathematical literature.

The works, which most impressed al-Samawal were those written by al-Karaji. Al-Samawal was satisfied with them and began to work out improvements for himself. His most famous treatise "Brilliant in algebra" was written when he was only nineteen years old. The treatise consists of four books: (1) "On premises, multiplication, division, and extraction of roots", (2) "On extraction of unknown quantities," (3) "On irrational magnitudes," and (4) "On classification of problems."

Al-Samawal's predecessors had begun to develop what has been called by historians today the "arithmetization of algebra." Al-Samawal was the first to give this development a precise description when he wrote that it was concerned "with operating on unknowns using all the arithmetical tools, in the same way as the arithmetician operates on the known."

Al-Samawal could not have described arithmetic operations on powers of the unknown without having developed an understanding of negative numbers. He wrote, "If we subtract a positive number from an empty power, the same negative number remains". In modern notation, it is: $0 - a = -a$. He continued, "If we subtract a negative number from an empty power, the same positive number remains". In modern notation, it is: $0 - (-a) = a$.

Multiplication of negative numbers was also understood by al-Samawal. He wrote, "The product of a negative number by a positive number is negative, and a negative number by a negative number is positive."

One of the most remarkable achievements appearing in Book 2 is al-Samawal's use of an early form of induction. Based on it, al-Samawal developed a method for summations of the first n natural numbers squared. In modern mathematical notation, his method is expressed as follows:

$$1^2 + 2^2 + 3^2 + \ldots + n^2 = n(n+1)(2n+1)/6$$

Let's check if this equation is correct. As an example, the numbers from 1 to 5 are taken in natural order.

The left-hand part of the equation gives:

$$1^2 + 2^2 + 3^2 + 4^2 + 5^2 = 1 + 4 + 9 + 16 + 25 = 55$$

The right-hand part of the equation gives:

$$5 \times (5+1)(2 \times 5 + 1)/6 = 5 \times 6 \times 11/6 = 330/6 = 55$$

Most of the works of al-Samawal have not survived, but some mathematical writings have survived. They contain work on fractions with the sum of fractions with numerators 1. This is an example:

$$\frac{80}{3\times7\times9\times10} = \frac{1}{3\times10} + \frac{1}{3\times7\times9} + \frac{1}{3\times9\times10}$$

Let check if this equation is correct.
Simplification of the left-hand part of the equation gives:

$$\frac{80}{3\times7\times9\times10} = \frac{80}{1890}$$

Simplification of the right-hand part of the equation gives:

$$\frac{1}{30} + \frac{1}{189} + \frac{1}{270} = \frac{1\times63+1\times10+1\times7}{1890} = \frac{80}{1890}$$

Conclusion: the left-hand part and the right-hand part of the equation are equal.

7.7 Nasir al-Din Tusi (February 17, 1201–June 25, 1274)

Khawaja Muhammad ibn Muhammad ibn Hasan Tusi, known as Nasir al-Din Tusi, was a Persian polymath, astronomer, biologist, chemist, mathematician, philosopher, physician, physicist, scientist, theologian, and the highest authority on religious laws (Ref 57).

Tusi was the first to write a work on trigonometry independently of astronomy. In his works trigonometry achieved the status of an independent brunch of pure mathematics distinct from astronomy, to which it had been linked for so long.

Tusi discovered the famous law of sines for plane triangles. This law is an equation relating the length of the sides of any shaped triangle to sines of its angles. According to the law:

$$\frac{a}{\sin A} = \frac{b}{\sin B} = \frac{c}{\sin C}$$

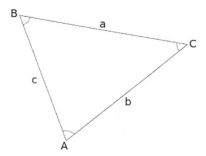

A triangle labelled with the components of the law of sines. Capital A, B, and C are the angles. Lower-case a, b, c are the sides opposite them: a opposite A, b opposite B, and c opposite C.

Tusi also stated the law of sines for spherical triangles, discovered the law of tangents for spherical triangles, and provided proofs for these laws.

Tusi established a reputation as an exceptional scholar. Tusi's prose writing, which number over 150 works, represent one of the largest collections by a single Islamic author. Writing in both Arabic and Persian, his works include the definitive <u>Arabic versions of the works of Euclid, Archimedes, Ptolemy, and Theodosius[5])</u>.

---

[5] Theodosius of Bithynia (c. 160 BC–c. 100 BC) was a Greek astronomer and mathematician who wrote the *Sphaerics*, a book on the geometry of the sphere.

Legacy of Tusi

A 60-km diameter lunar crater located on the southern hemisphere of the moon is named after him as "Nasireddin." A minor planet discovered by Soviet astronomer Nikolai Stepanovich Chernykh in 1979 was named 10269 Tusi.

The K. N. Toosi University of Technology in Iran and Observatory of Shamakhy in the Republic of Azerbaijan are also named after him.

In February 2013, Google celebrated his 812th birthday with a doodle, which was accessible in its websites with Arabic language calling *al-farsi* (the Persian).

His birthday is also celebrated as Engineer's Day in Iran.

7.8 Jamshid al-Kashi (1380–June 22, 1429)

Ghiyath al-Din Jamshid Masud al-Kashi (or al-Kashani), known as Jamshid al-Kashi (Ref 58) was a Persian astronomer and mathematician. Much of his work was not brought to Europe, and much, even the extant work remains unpublished in any form.

Al-Kashi was one of the best mathematicians in the history of the Persian Empire.

The period of Shah Rokh power (1409–1417) became one of many scholarly accomplishments. This was the perfect environment for al-Kashi to begin his career as one of the world's greatest mathematicians. He was the first to provide a precise statement of the law of cosines in a form suitable for triangulation. The law of cosines used for calculating one side of a triangle when the angle opposite and the other two sides are known.

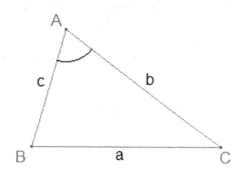

$$a^2 = b^2 + c^2 - 2bc \times \cos A$$
$$b^2 = c^2 + a^2 - 2ca \times \cos B$$
$$c^2 = a^2 + b^2 - 2ab \times \cos C$$

Al-Kashi in *The Treatise on the Chord and Sine* computed sin 1° to accuracy, which was the most accurate approximation in his time. In order to determine sin 1°, he discovered the following formula:

$$\sin 3\alpha = 3 \sin \alpha - 4 \sin^3 \alpha$$

Let's check this formula for accuracy. Calculations where performed using calculator *TI-30X SOLAR*.

Let $\alpha = 1°$. The left-hand part of the equation is:

$$\sin 3 = \underline{0.052335956}$$

The right-hand part of the equation is:

$3 \sin 1 - 4 \sin^3 1 =$
$3 \times 0.017452406 - 4 \times 0.017452406^3 =$
$0.052357219 - 4 \times 0.000005316 =$
$0.052357219 - 0.000021263 = \underline{0.052335956}$

As we can see, the left-hand part and the right-hand part of the equation are equal, so the formula is accurate.

In algebra and numerical analysis, al-Kashi developed an iterative method for solving cubic equations, which was not discovered in Europe until centuries later.

In 1424, al-Kashi correctly computed 2π to 16 decimal places of accuracy. This was far more accurate than the estimates obtained by Greek mathematics (3 decimal places by Ptolemy), or Indian mathematics (11 decimal places by Madhava of Sangamagrama). The accuracy of al-Kashi's estimate was not surpassed until 1596, when Ludolph van Ceulen[6]) computed 20 decimal places of π.

Al-Kashi was still working on his book, called "Risala al-watar wa'l-jaib" meaning "The Treatise on the Chord and Sine," when he died in 1429.

---

[6] Ludolph van Ceulen (1540–1610) was a German-Dutch mathematician.

CHAPTER 8

MEDIEVAL EUROPEAN MATHEMATICS

This period lasted from the 5th to 14th centuries. Medieval European interest in mathematics (Ref 2) was limited to understanding the created order of nature, justified by Plato's *Timaeus* (one of his dialogues) and the biblical passage in the Book of Wisdom[1]) that God had *ordered all things in measure, number, and weight.*

Boethius (c. 480–524 AD) was a Roman senator, consul, philosopher, and one of the most senior administrative officials. He composed *Consolation of Philosophy*, a philosophical treatise on fortune, death, and other issues, which became one of the most popular and influential works of the Middle Ages. He provided a place for mathematics in the curriculum under the term *quadrivium* (four subjects), which described the study of arithmetic, geometry, astronomy, and music.

Boethius wrote translated from the Greek *Introduction to Arithmetic* and a series of excerpts from Euclid's *Elements*. His works were theoretical, rather than practical, and were the basis of mathematical study until the recovery of Greek and Arabic mathematical works.

In the 12th century, European scholars traveled to Spain and Sicily seeking scientific Arabic texts, including al-Khwarizmi's *The Compendious Book on Calculation by Completion and Balancing* and

---

[1] The Book of Wisdom or Wisdom of Solomon is one of the books of the Bible.

the complete text of Euclid's *Elements* translated in Latin. These new sources sparked a renewal of mathematics.

8.1 Leonardo Fibonacci (c. 1175–c. 1250)

Leonardo Fibonacci (Ref 59) was an Italian mathematician, considered to be the most-talented Western mathematician of the middle ages. Fibonacci is best known for the sequence of numbers, which he developed.

Leonardo was born into a wealthy family. His father was the consul for Pisa and directed a trading post in Bugia (now Béjaïa, Algeria) a port in North Africa. As a young boy, he travelled with his father. In Bugia he studied the Hindu–Arabic numeral system.[2])

Fibonacci popularized the Hindu–Arabic numeral system to the Western World primarily through his composition in 1202 of *Liber Abaci* (*Book of Calculation*). He also introduced Europe to the sequence of Fibonacci numbers, which he used as an example in *Liber Abaci*.

Liber Abaci posed and solved a problem involving the growth of a population of rabbits based on idealized assumptions. The solution,

---

[2] It is a positional decimal numeral system that is the most common system for the symbolic representation of numbers in the world.

generation by generation, was a sequence of numbers later known as Fibonacci numbers.

In the Fibonacci sequence of numbers, each number is the sum of the previous two numbers. Fibonacci began the sequence not with 0, 1, 1, 2, 3... as modern mathematicians do, but with 1, 1, 2, 3... He carried the calculation up to the thirteenth place (fourteenth in modern counting), that is 233, though another manuscript carries it to the next place: 1, 1, 2, 3, 5, 8, 13, 21, 34, 55, 89, 144, 233, and 377. Fibonacci did not speak about the golden ratio[3] as the limit of the ratio of <u>consecutive numbers in this sequence.</u>

In mathematics, two quantities are in the golden ratio if their ratio is the same as the ratio of their sum to the larger of the two quantities. The figure below illustrates the geometric relationship (Ref 60).

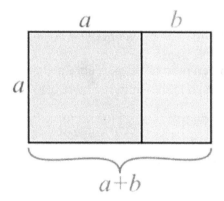

A golden rectangle (in pink) with shorter side b and longer side a, when placed adjacent to a square with sides of length a, will produce a similar golden rectangle with longer side $a + b$ and shorter side a. Algebraic notation for quantities b and a with $a > b > 0$, is expressed by the formula:

$$\varphi = \frac{a+b}{a}$$

---

[3] The golden ratio is also called the golden mean or golden section. Other names include golden proportion, golden cut, and golden number.

where φ represents the golden ratio. Its value is:

$$\varphi = \frac{1+\sqrt{5}}{2} = 1.618033989$$

Mathematicians since Euclid have studied the properties of the golden ratio, including its appearance in the dimensions of a regular pentagon (it is a five-sided polygon, in which the sum of the internal angles is 540°) and in a golden rectangle, which may be cut into a square and a smaller rectangle with the same aspect ratio (it is an image describes the proportional relationship between its width and its height. It is commonly expressed as two numbers separated by colon, as in 16:9).

The golden ratio has also been used to analyze the proportions of natural objects as well as man-made systems such as financial markets, in some cases based on dubious fits to data. The golden ratio appears in some patterns in nature, including the spiral arrangement of leaves and other plant parts.

Value of the golden ratio can also be obtained through the Fibonacci sequence of numbers. It is shown in the table below.

Place in the sequence	Fibonacci number	Ratio of places	Value of ratio for related numbers	Deviation from the golden ratio 1.618033989
12	89	13/12	1.617977528	0.000056461
13	144	14/13	1.618055556	-0.000021567
14	233	15/14	1.618025751	0.000008238
15	377	16/15	1.618037135	-0.000003146
16	610	17/16	1.618032787	0.000001202
17	987	18/17	1.618034448	-0.000000459
18	1597	19/18	1.618033813	0.000000176
19	2584	20/19	1.618034056	-0.000000067
20	4181	21/20	1.618033963	0.000000026
21	6765	22/21	1.618033999	-0.000000010
22	10946	23/22	1.618033985	0.000000004

23	17711	24/23	1.618033990	-0.000000001
24	28657	25/24	1.618033988	0.000000001
25	46368	26/25	1.618033989	0.000000000
26	75025	27/26	1.618033989	0.000000000
27	121393	28/27	1.618033989	0.000000000
28	196418			

Some of the greatest mathematical minds of all ages, from Pythagoras and Euclid in ancient Greece, through the medieval Italian mathematician Leonardo Fibonacci to present-day scientific figures, have spent endless hours over this simple ratio and its properties. But the fascination with the Golden Ratio is not confined just to mathematicians. Biologists, artists, musicians, historians, architects, psychologists, and even mystics have pondered and debated the basis of its ubiquity and appeal. In fact, it is probably fair to say that the Golden Ratio has inspired thinkers of all disciplines like no other number in the history of mathematics.

The Fibonacci spiral approximates the golden spiral using quarter-circle arcs inscribed in squares of integer Fibonacci-number side, shown for square sizes 1, 1, 2, 3, 5, 8, 13, 21, and 34 (Ref 61).

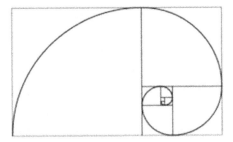

There are many mathematical concepts named after Fibonacci, because of a connection to the Fibonacci numbers, such as the Brahmagupta-Fibonacci identity, Fibonacci search technique, and others.

In algebra, the Brahmagupta-Fibonacci identity or simply Fibonacci's identity says that the product of two sums each of two squares is itself a sum of two squares. In other words, the set of all sums of two squares is closed under multiplication. Specifically:

$$(a^2 + b^2)(c^2 + d^2) = (ac - bd)^2 + (ad + bc)^2 \qquad (1)$$
$$(a^2 + b^2)(c^2 + d^2) = (ac + bd)^2 + (ad - bc)^2 \qquad (2)$$

The following calculations are taken to show that these equations are accurate. The left-hand parts of both equations are identical. Let $a = 2, b = 3, c = 4, d = 5$. Then:

$$(a^2 + b^2)(c^2 + d^2) = (2^2 + 3^2)(4^2 + 5^2) = (4 + 9)(16 + 25) = 13 \times 41 = \underline{533}.$$

The right-hand part of the equation (1) is:

$$(ac - bd)^2 + (ad + bc)^2 = (2 \times 4 - 3 \times 5)^2 + (2 \times 5 + 3 \times 4)^2 = (-7)^2 + 22^2 = 49 + 484 = \underline{533}.$$

The right-hand part of the equation (2) is:

$$(ac + bd)^2 + (ad - bc)^2 = (2 \times 4 + 3 \times 5)^2 + (2 \times 5 - 3 \times 4)^2 = 23^2 + (-2)^2 = 529 + 4 = \underline{533}.$$

In computer science, the Fibonacci search technique is a method of searching a sorted array using a "divide and conquer" algorithm that narrows down possible locations with the aid of Fibonacci sequence. The name "Fibonacci sequence" was first used by Édouard Lucas (1842–1891). He was a French mathematician, specialized in the number theory.

Beyond mathematics, namesake of Fibonacci includes the asteroid "6765 Fibonacci".

The date of Fibonacci's death is not known, but it has been estimated to be between 1240 and 1250, most likely in Pisa.

8.2 Thomas Bradwardine (c. 1290–1349)

Thomas Bradwardine (Ref 62) was an English scholar, mathematician, physicist, courtier, cleric and Archbishop of Canterbury. As a celebrated scholastic philosopher, and doctor of theology, he is often called "the Profound Doctor."

Bradwardine was one of the Oxford Calculators (a group of thinkers, associated with Merton College) studying mechanics. They distinguished kinematics from dynamics, emphasizing kinematics, and investigating instantaneous velocity. They first formulated the mean speed theorem: "A body moving with constant velocity travels the same distance as an accelerated body in the same time if its velocity is half the final speed of the accelerated body." They also demonstrated this theorem long before Galileo, who is credited with it.

Thomas Bradwardine proposed that speed (V) increases in arithmetic proportion as the ratio of force (F) to resistance (R) increases in geometric proportion. Bradwardine expressed this by a series of specific examples, but although the logarithm had not yet been conceived. In modern mathematical notation, his conclusion can be expressed as:

$$V = \log (F/R).$$

At the time of Bradwardine, the logarithms didn't exist. They were invented three hundred years later by a Scottish mathematician, physicist, and astronomer John Napier (1550–1617). He published his logarithms in 1614.

8.3 William Heytesbury (1313–1372 / 1373)

William of Heytesbury, or William Heytesbury (Ref 63) was an English philosopher and logician, best known as one of the Oxford Calculators[4] of Merton College[5]) in Oxford, England.

William Heytesbury (Ref 2), lacking differential calculus and the concept of limits, proposed to measure instantaneous speed "by the path that would be described by a body, if it were moved uniformly at the same degree of speed with which it is moved in that given instant". He mathematically determined the distance covered by a body undergoing uniformly accelerated motion[6], stating that "a moving body uniformly acquiring or losing that increment of speed will traverse in some given time a distance completely equal to that, which it would traverse if it were moving continuously through the same time with the mean degree of speed."

For mathematics and mechanics (Ref 63, text in Russian was translated in English by Edmund Isakov) of particular interest are Heytesbury's foundations of the theory developed by the scientists of

---

[4] The Oxford Calculators were a group of 14th-century thinkers, almost all associated with Merton College. For this reason they were dubbed "The Merton School". These men took a strikingly logical-mathematical approach to philosophical problems. The key "calculators," were Thomas Bradwardine, William Heytesbury, Richard Swineshead and John Dumbleton.

[5] Merton College (in full: The House or College of Scholars of Merton in the University of Oxford).

[6] In today's mathematics such distance is calculated by integration.

Merton College on the uniform movement, which was contrasted with the movement of the uneven ("differential").

As applied to the uneven motion, Heytesbury distinguishes its subclass—the uniform motion. He gives a completely clear definition of the uniform motion, asserting, "Every motion is uniformly accelerated if, for any equal part of the time, it acquires an equal increment in velocity."

Heytesbury introduced into mechanics the notion of instantaneous speed: "The speed at any given time will be determined by the way that would be described moving point if for a certain period of time it would move uniformly with that degree of speed, with which it moved at this moment, no matter what time it was indicated."

For the case of alternating motion, Heytesbury formulated and proved the so-called theorem on the mean degree of velocity. The theorem asserts that the path traveled by the body for some time with the same alternating motion is equal to the path traveled by the body during the same time with uniform motion with a speed equal to the arithmetic average of the maximum and minimum values of the velocity in the alternating motion. In modern mathematical notation, distance traveled can be expressed as:

$$S = (V_0 + V)/2 \times T$$

Where S is a distance traveled, V_0 and V are initial and final velocities in uniformly moving, and T is a time of movement.

8.4 Nicole Oresme (c. 1320/5–1382)

Nicole Oresme, also known as Nicholas Oresme, or Nicolas d'Oresme (Ref 64), was a significant philosopher of the later Medieval Ages. He wrote influential works on economics, mathematics, physics, astrology and astronomy, philosophy, and theology; was Bishop of Lisieux[7], a translator, a counselor of King Charles V[8] of <u>France, and probably one of the most original thinkers of the 14th century Europe.</u>

Nicole Oresme was born in the village in the vicinity of Caen, Normandy in France. Nothing is known concerning his family. He was studying at the University of Paris and received a degree Master of Arts (Latin: *Magister Artium*).

In 1342, Oresme was already a Regent Master (a title conferred in the medieval universities meant the right to teach).

In 1348, Oresme was studying theology (it is a nature of the divine) at the University of Paris. In 1356, he received his doctorate and became grand master of the College of Navarre (it was one of the colleges of the University of Paris).

---

[7] Lisieux is a town in the Normandy region in northwestern France.
[8] King Charles V (1338–1380) called the Wise, was a monarch of the House of Valois who ruled as King of France from 1364 to his death.

In 1364, Oresme was appointed dean of the Cathedral of Rouen (a city on the River Seine in the north of France). In 1369, he began translating the works by Aristotle[9]) at the request of Charles V, the King of France.

In 1371, Charles V granted Oresme a pension. In 1377, with royal support, he was appointed bishop of Lisieux (department in the Normandy region in France).

Oresme's important contributions to mathematics were:

- The idea of visualizing *latitude* and *longitude* by plane figures, approaching what we would now call rectangular coordinates before Descartes (1596–1650).
- The first proof of the divergence[10] of the harmonic series[11], something that was only replicated by the Bernoulli brothers, Jacob (1654–1705) and Johann (1667–1748).

$$\sum_{n=1}^{\infty} \frac{1}{n} = 1 + \frac{1}{2} + \frac{1}{3} + \frac{1}{4} + \frac{1}{5} + \cdots$$

Oresme was the first, who developed the proof of the divergence of the harmonic series. He noted that for any n that is a power of 2, there are $n/2 - 1$ terms in the series between $2/n$ and $1/n$. Each of these terms is at least $1/n$, and since there are $n/2$ of them, they sum to at least $1/2$.

For instance, there is one term $1/2$, then two terms $1/3 + 1/4$ that together sum to at least $1/2$, then four terms $1/5 + 1/6 + 1/7 + 1/8$ that also sum to at least of $1/2$, and so on.

Let's calculate the sums of two, four, and six terms.

Two terms:

$$1/3 + 1/4 = \frac{1 \times 4 + 1 \times 3}{3 \times 4} = \frac{7}{12} = 0.5833$$

---

[9] Aristotle (384–322 BC) was an ancient Greek philosopher and scientist.
[10] In mathematics, a divergent series is an infinite series that is not convergent, meaning that the infinite sequence of the partial sums of the series does not have a finite limit.
[11] In mathematics, the harmonic series is the divergent infinite series:

Four terms:

$$1/5 + 1/6 + 1/7 + 1/8 = \frac{336+280+240+210}{5\times6\times7\times8} = \frac{1066}{1680} = 0.6345$$

Six terms:

$$1/9 + 1/10 + 1/11 + 1/12 + 1/13 + 1/14 =$$

$$\frac{240240+216216+196560+180180+166320+154440}{9\times10\times11\times12\times13\times14} = \frac{1153956}{2162160} = 0.5337$$

The sum of each term, shown above, is greater than $1/2$.

Thus, the series must be greater than the series $1 + 1/2 + 1/2 + 1/2 + ...$, which does not have a finite limit. This proves that the harmonic series must be divergent.

Oresme was the first mathematician to prove this fact, and (after his proof was lost) it was not proven again until the 17th century by Pietro Mengoli (1628–1686), the Italian mathematician.

Oresme also worked on fractional powers and the notion of probability over infinite sequences, ideas which would not be further developed for the next three and five centuries, respectively.

The crater "Oresme" on the far side of the Moon was named after him by the International Astronomical Union in 1970.

CHAPTER 9

RENAISSANCE MATHEMATICS

During the Renaissance (Ref 2), a period in Europe from the 14th to the 17th century, the development of mathematics and of accounting were associated. While there is no direct relationship between algebra and accounting, the teaching of the subjects, and the books published often intended for the children of merchants. They were sent to reckoning schools (in Flanders[1] and Germany) or abacus schools in Italy, where they learned the skills useful for trade and commerce. Probably, there was no need for algebra in performing bookkeeping operations, but for bartering operations or the calculation of compound interest, a basic knowledge of arithmetic <u>was mandatory and knowledge of algebra was very useful.</u>

---

[1] Flanders today normally refers to the Dutch-speaking northern portion of Belgium.

9.1 Piero Francesca (c. 1415–1492)

A self-portrait, detail from
The Resurrection

Piero Francesca (Ref 65) was an Italian painter, mathematician, and geometer. His painting was characterized by its use of geometric forms and perspective[2]).

Piero's deep interest in the theoretical study of perspective and his contemplative approach to his paintings are apparent in all his work. Three treatises written by Piero are known to modern mathematicians: *Abacus Treatise, Short Book on the Five Regular Solids,* and *On Perspective for Painting*. The subjects covered in these writings include arithmetic, algebra, geometry and innovative work in both solid geometry and perspective. Much of Piero's work was later absorbed into the writing of others, notably Luca Pacioli. Piero's work on solid geometry appears in Pacioli's *On the Divine Proportion,* a work illustrated by Leonardo da Vinci.

In the late 1450s, Piero copied and illustrated the following works of *Archimedes: On the Sphere and the Cylinder, On the Measurement of*

---

[2] Perspective in the graphic arts is an approximate representation on a flat surface an image as it is seen by the eye. The two most characteristic features of perspective are that objects are smaller as their distance from the observer increases.

the Circle, On Conoids and Spheroids, On Spirals, On the Equilibrium of Planes, On the Quadrature of the Parabola,* and *The Sand Reckoner*. The manuscript consists of 82 folio leaves. It's held in the collection of the Riccardian Library[3]).

9.2 Luca Pacioli (c. 1447–1517)

The portrait reads: "Quantity is power".

Fra Luca Bartolomeo de Pacioli, also known as Luca Pacioli (Ref 66), was an Italian mathematician, Franciscan friar[4], collaborator with Leonardo da Vinci, and productive contributor to the field now known as accounting. He is referred to as "The Father of Accounting and Bookkeeping."

Pacioli published several works on mathematics, including:

- *Summary of arithmetic, geometry, proportions and proportionality* (published in 1494), a textbook for use in the schools of

---

[3] The Riccardian Library is a library in Florence, Italy. The library is located in the Palazzo Medici Riccardi.
[4] A friar is a member of one of the Christian religious orders founded since the twelfth or thirteenth century. Religious orders have adopted a lifestyle of poverty, travelling, and living in urban areas for purposes of preaching, evangelization, and ministry, especially to the poor.

Northern Italy. It was a synthesis of the mathematical knowledge of his time and contained the first printed work on algebra written in the vernacular (i.e. the spoken language of the day). It is also notable for including the first published description of the bookkeeping method that Venetian merchants used during the Italian Renaissance.
- *Geometry*, a Latin translation of Euclid's *Elements* (published in 1509).
- *On the Divine Proportion* illustrated by Leonardo da Vinci and published in Venice in 1509. The subject was mathematical and artistic proportion, especially the mathematics of the golden ratio[5] and its application in architecture. Leonardo da Vinci drew the illustrations of the regular solids in *On the Divine Proportion* while <u>he lived with and took mathematics lessons from Pacioli.</u>

Leonardo da Vinci's drawings are probably the first illustrations of a skeleton solids, which allowed an easy distinction between front and back. This one was the first printed illustration of a rhombicuboctahedron in 1509.

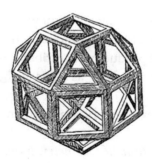

In geometry, the rhombicuboctahedron is an Archimedean solid with eight triangular and eighteen square faces.

---

[5] Golden ratio was described earlier (Section 8.1).

9.3 Scipione Del Ferro (1465–1526)

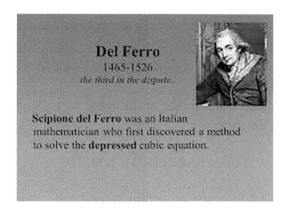

Scipione Del Ferro was an Italian mathematician (Ref 67), who first discovered a method to solve a depressed cubic equation. Such equation has only 1 root and 2 imaginary roots.

There are no surviving scripts from Del Ferro. This is in large part due to his resistance to communicating his works. Instead of publishing his ideas, he would only show them to a small, select group of friends and students.

Del Ferro graduated from the University of Bologna in 1496 and was appointed as a lecturer in arithmetic and geometry (Ref 68). He retained this post for the rest of his life. After years of efforts, Del Ferro managed to develop a formula for solving cubic equation of the form: $x^3 + ax = b$, where $a > 0$ and $b > 0$. The formula is:

$$x = \sqrt[3]{\frac{b}{2} + \sqrt{\left(\frac{b}{2}\right)^2 + \left(\frac{a}{3}\right)^3}} + \sqrt[3]{\frac{b}{2} - \sqrt{\left(\frac{b}{2}\right)^2 + \left(\frac{a}{3}\right)^3}}$$

Case for $b > a$. Let $b = 10$ and $a = 9$

1) Calculate of data in the inner roots at the right-hand part of the formula:

$(b/2)^2 + (a/3)^3$

$(10/2)^2 = 5^2 = 25$
$(9/3)^3 = 3^3 = 27$
$25 + 27 = 52$
$\sqrt{52} = \underline{7.211103}$

As you can see, the first numbers of cube roots are $b/2 = 10/2 = 5$

2) Continuing calculate the data of the cubic roots. To do it, we used the Cube Root Calculator.

$$\sqrt[3]{5 + 7.211103} = \sqrt[3]{12.211103} = 2.302776$$

3) Continue to calculate the data of the second cubic root

$$\sqrt[3]{5 + 7.211103} = \sqrt[3]{-2.211103} = -1.302776$$

4) The sum of results of the above cubic roots

$$x = 2.302776 + (-1.302776) = 1$$

We are checking the equation $x^3 + ax = b$ for its accurate. Let $x = 1$, $a = 9$, and $b = 10$. The result is:

$$1^3 + 9 \times 1 = 10$$

The accuracy of the Del Ferro's formula for data of Case $b > a$ is 100%

<u>Case for $b < a$.</u> Let's $b = 8$ and $a = 12$

1) Calculate the data of the inner roots at the right-hand part of the formula:

$(b/2)^2 + (a/3)^3$
$(8/2)^2 = 4^2 = 16$
$(12/3)^3 = 4^3 = 64$

$$16 + 64 = 80$$
$$\sqrt{80} = 8.944272$$

As you can see, the first numbers of cube roots are $b/2 = 8/2 = 4$

2) Continuing to calculate the data in the cubic roots. To do it, we use the Cube Root Calculator

$$\sqrt[3]{4 + 8.944272} = \sqrt[3]{12.944272} = 2.34797$$

3) Continue to calculate the data of the second cubic root

$$\sqrt[3]{4 - 8.944272} = \sqrt[3]{-4.944272} = -1.703599291$$

4) The sum of results of the above cubic roots

$$x = 2.34797 + (-1.7036) = 0.644371$$

We are checking the equation $x^3 + ax = b$ for its accurate. Let $x = 0.644371$, $a = 12$, and $b = 8$. The result is:

$$0.644371^3 + 12 \times 0.644371 = 0.267552 + 7.732452 = 8$$

The accuracy of Del Ferro's equation for data of Case $b < a$ is 100%.

Del Ferro's role in the history of mathematics is an important one, and he deserves great credit for solving one of the outstanding ancient problems of mathematics—cubic equations (Ref 69).

Del Ferro also made other important contributions to the rationalization of fractions with denominators containing sums of cube roots. He also investigated geometry problems with a compass set at a fixed angle, but little is known about his work in this area.

9.4 Niccolò Fontana Tartaglia (1500–1557)

Niccolò Fontana Tartaglia (Ref 70) was an Italian mathematician, engineer (designing fortifications), a surveyor, and a bookkeeper. He published many books, including the first Italian translations of Archimedes and Euclid, and an acclaimed compilation of mathematics. Tartaglia was the first to apply mathematics to the investigation of the paths of cannonballs, known as ballistics.

In 1512, the French troops invaded Brescia (a city in the region of Lombardy in northern Italy) during the war against Venice. Niccolò was wounded by the French soldier, who sliced Niccolò's jaw and palate with a saber. His mother nursed him back to health, but the young boy didn't recover his normal speech. He took "Tartaglia" (in English—"stammerer") as a nickname, which referred to his inability to talk clearly as a result of terrible wounds. After this, he had never shaved, and grew a beard to camouflage his scars.

There is a story that Tartaglia learned only half the alphabet from a private tutor before funds ran out, and he had to learn the rest by himself. Be that as it may, he was essentially self-taught. He and his contemporaries, working outside the academies, were responsible for the spread of classical works in modern languages among the educated middle class.

Tartaglia is best known today for his conflicts with Gerolamo Cardano[6].

Cardano cajoled Tartaglia into revealing his solution to the cubic equations, by promising not to publish them. Tartaglia divulged the secrets of the solutions of three different forms of the cubic equation in verse. Several years later, Cardano happened to see unpublished work by Scipione Del Ferro who independently came up with the same solution as Tartaglia. As the unpublished work was dated before Tartaglia's, Cardano decided his promise could be broken and included Tartaglia's solution in his next publication. Even though Cardano credited his discovery, Tartaglia was extremely upset, and a famous public challenge match resulted between himself and Cardano's student, Ludovico Ferrari[7].

Mathematical historians now credit both Cardano and Tartaglia with the formula to solve cubic equations, referring to it as the "Cardano–Tartaglia formula."

Tartaglia's best known work is *General Treaty of Numbers and Other Measures* published in Venice 1556–1560. This has been called the best treatise on arithmetic that appeared in the sixteenth century. Not only does Tartaglia have complete discussions of numerical operations and the commercial rules used by Italian arithmeticians in this work, but he also discusses the life of the people, the customs of merchants and the efforts made to improve arithmetic in the 16th century.

Tartaglia liked arithmetic puzzles. This one was his famous (Ref 71): *A man dies leaving 17 horses to be divided among his heirs in the ratio 1 / 2, 1 / 3, and 1 / 9. How can this be done?* Tartaglia's solution

---

[6] Gerolamo Cardano was an Italian polymath and the one of the most influential mathematicians of the Renaissance (detailed information on his contributions to mathematics is provided in Section 9.5).

[7] Ludovico Ferrari (1522–1565) was an Italian mathematician. He aided Cardano on his solutions for quadratic equations and cubic equations. Ferrari was responsible for the solution of quartic equations that Cardano published. While in his teens, Ferrari obtained a prestigious post to teach mathematics. He retired at the age of 42 and moved to Bologna to take up a professorship of mathematics at the University of Bologna. In 1565, he died of white arsenic poisoning, administered by his sister.

involved borrowing an extra horse for calculating the distribution and then politely returning it after the calculation. As a matter of fact, one extra horse needs to only multiply 18 by the ratios:

$18 \times 1/2 = 9$ (9 horses for the first heir)
$18 \times 1/3 = 6$ (6 horses for the second heir)
$18 \times 1/9 = 2$ (2 horses for the third heir)
$9 + 6 + 2 = 17$

9.5 Gerolamo Cardano (1501–1576)

Girolamo Cardano
(1501-1576)

Gerolamo Cardano (Ref 72) was an Italian polymath, whose interests and knowledge ranged from being a mathematician, physician, biologist, physicist, chemist, astrologer, astronomer, philosopher, writer, and renowned mechanic. He was one of the most influential mathematicians of the Renaissance, and was one of the key figures in the foundation of probability and the earliest introducer of the binomial coefficients and the binomial theorem in the western world. He wrote more than 200 works on science. Cardano was the first mathematician to make systematic use of numbers less than zero.

In 1545, Cardano published in Latin his book on algebra *Ars Magna* (*The great Art*, in English). He acknowledged his student Ludovico

Ferrari for solution of the quartic equation[8]) and Scipione Del Ferro for solution of the cubic equation.

Cardano acknowledged the existence of what are now called imaginary numbers[9]). He didn't understand their properties, described for the first time by his Italian contemporary mathematician Bombelli[10]).

Cardano's "Book on Games of Chance" (written in 1564, but not published until 1663) contains the first systematic treatment of probability. Cardano made several contributions to hydrodynamics and held that perpetual motion is impossible, except in celestial bodies. He published two encyclopedias of natural science, which contain a wide variety of inventions, facts, and occult superstitions.

Cardano invented and described several mechanical devices, among which combination lock, gimbal[11]) consisting of three concentric rings, and Cardan shaft[12]).

May be used to allow an object mounted on the innermost gimbal to remain independent of the rotation of its support. For example, on the ship, the gyroscopes, shipboard compasses, and even drink holder use gimbals to keep them upright with respect to the horizon despite the ship's pitching and rolling.

---

[8] A quartic equation, or equation of the fourth degree, is an equation that equates a quartic polynomial to zero of the form $ax^4 + bx^3 + cx^2 + dx + e = 0$, where $a \neq 0$

[9] An imaginary number is a complex number that can be written as a real number multiplied by the imaginary unit i, which is defined by its property $i^2 = -1$.

[10] Rafael Bombelli (1526–1572) was an Italian mathematician (detailed information on his contributions to mathematics is provided in Section 9.6).

[11] A gimbal is a pivoted support that allows the rotation of an object about a single axis. A set of three gimbals, one mounted on the other with orthogonal pivot axes,

[12] Cardan shaft (a drive shaft named after Cardano) is a mechanical component for transmitting torque and rotation, used to connect other components of a drive train that cannot be connected directly, because of distance or the need to allow for relative movement between them.

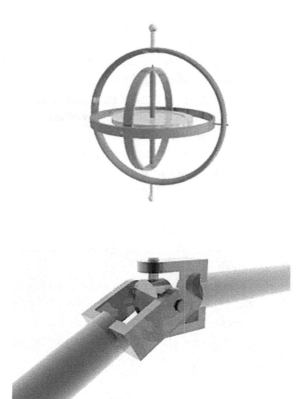

Cardan shaft is a priceless invention. It is used in myriad of vehicles to this day.

9.6 Rafael Bombelli (1526–1572)

Rafael Bombelli (Ref 73) was an Italian mathematician. He felt that none of the works on algebra by the leading mathematicians of his day provided precise and complete analysis of the subject.

Instead of another complex treatise that only mathematicians could comprehend, Rafael decided to write a book on algebra that could be understood by anyone. His text would be self-contained and easily read by those without higher education. He was the one who finally managed to address the problem with imaginary numbers.

An imaginary number (Ref 74) is a complex number that can be written as a real number multiplied by the imaginary unit i, which is defined by its property $i^2 = -1$. In modern mathematical notation, $i = \sqrt{-1}$.

In 1572, Rafael Bombelli provided the rules for multiplication of complex numbers. At the time, such numbers were poorly understood and regarded by some mathematicians as fictitious or useless, much as zero and the negative numbers. Some mathematicians were against to adopt the use of imaginary numbers. He gave a special name to square roots of negative numbers, instead of just trying to deal with them as regular radicals like other mathematicians did. He called the imaginary number i "plus of minus" and used "minus of minus" for $-i$.

Bombelli had the foresight to see that imaginary numbers were crucial and necessary to solving quartic and cubic equations. At the time, people cared about complex numbers only as tools to solve practical equations.

Bombelli is generally regarded as the inventor of complex numbers, as no one before him had made rules for dealing with such numbers, and no one believed that working with imaginary numbers would have useful results (Ref 74).

9.7 François Viète (1540–1603)

François Viète, in Latin Franciscus Vieta, (Ref 75) was a French mathematician whose work on new algebra[13] was an important step towards modern algebra, due to <u>its innovative use of letters as parameters in equations.</u>

Viète was the first mathematician who introduced notations for the problem (and not just for the unknowns). As a result, his algebra was no longer limited to the statement of rules, but relied on an efficient computer algebra, in which the operations act on the letters and the results can be obtained at the end of the calculations by a simple replacement. This approach, which is the heart of contemporary algebraic method,

---

[13] The new algebra or symbolic analysis is a formalization of algebra promoted by François Viète in 1591 and by his successors after 1603. It marks the beginning of the algebraic formalization in the late sixteenth and the early seventeenth centuries.

was a fundamental step in the development of mathematics. With that, Vieta marked the end of medieval algebra and opened the modern period.

Viète created many innovations, such as the binomial formulas, which would be taken later by Pascal and Newton, and the link between the roots and coefficients of a polynomial. One of Viète's formulas applied to quadratic and cubic polynomial. For the quadratic equation such as $ax^2 + bx + c = 0$, roots x_1 and x_2 satisfy:

$$x_1 + x_2 = -b/a$$
$$x_1 \times x_2 = c/a$$

For the cubic equation such as $ax^3 + bx^2 + cx + d = 0$, roots x_1, x_2, and x_3 satisfy:

$$x_1 + x_2 + x_3 = -b/a$$
$$x_1 \times x_2 + x_1 \times x_3 + x_2 \times x_3 = c/a$$
$$x_1 \times x_2 \times x_3 = -d/a$$

As an example, let solve the quadratic equation $x^2 + 3x - 4 = 0$ by using the formula:

$$x_1, x_2 = \frac{-b \pm \sqrt{b^2 - 4ac}}{2a}$$

where $a = 1$, $b = 3$, and $c = 4$

$$x_1 = \frac{-3 + \sqrt{3^2 - 4(1)(-4)}}{2(1)} = \frac{-3 + \sqrt{25}}{2} = \frac{2}{2} = 1$$

$$x_2 = \frac{-3 - \sqrt{3^2 - 4(1)(-4)}}{2(1)} = \frac{-3 - \sqrt{25}}{2} = \frac{-8}{2} = -4$$

$$x_1 + x_2 = -b/a = -3/1 = -3$$
$$x_1 + x_2 = 1 - 4 = -3$$
$$x_1 \times x_2 = c/a = -4/1 = -4$$

Another of Viète's formula representing the mathematical constant π is:

$$\frac{2}{\pi} = \frac{\sqrt{2}}{2} \times \frac{\sqrt{2+\sqrt{2}}}{2} \times \frac{\sqrt{2+\sqrt{2+\sqrt{2}}}}{2} \cdots$$

The left-hand part of this formula expressed in modern mathematical notation, is:

$$\frac{2}{\pi} = 2/3.141592654 = \underline{0.636619772}$$

Let's calculate the right-hand parts of the formula, which contains three factors. We used the calculator *TI-30X SOLAR* TEXAS INSTRUMENTS.

The first factor is: $\frac{\sqrt{2}}{2} = 1.414213562/2 = 0.707106781$

The second factor is: $\frac{\sqrt{2+\sqrt{2}}}{2} = 0.923879533$

The third factor is: $\frac{\sqrt{2+\sqrt{2+\sqrt{2}}}}{2} = 0.98078528$

The product of these three factors is:

$0.707106781 \times 0.923879533 \times 0.98078528 = \underline{0.640728862}$

The value of the left-hand part of the formula (0.636619772) is less than the value of the right-hand part of the formula (0.640728862) by:

$0.640728862 - 0.636619772 = 0.00410909$, or $\approx 0.65\%$.

Let's see how the fourth factor affects the right-hand part of the formula.

The fourth factor is: $\frac{\sqrt{2+\sqrt{2+\sqrt{2+\sqrt{2}}}}}{2} = 0.995184727$

The product of all four factors is:

$0.640728862 \times 0.995184727 = 0.637643578$

The value of the left-hand part of the formula (0.636619772) is less than the value of the right-hand part of the formula (0.637643578) by:

0.637643578 − 0.636619772 = 0.001023806, or ≈ 0.16%

Vieta was the first mathematician who introduced notations for the problem (and not just for the unknowns). As a result, his algebra was no longer limited to the statement of rules, but relied on an efficient computer algebra, in which the operations act on the letters and the results can be obtained at the end of the calculations by a simple replacement. This approach, which is the heart of contemporary algebraic method, was a fundamental step in the development of mathematics. With this, Vieta marked the end of medieval algebra and opened the modern period.

9.8 Bartholomeo Pitiscus (1561–1613)

Bartholomeo Pitiscus (Ref 76) was a German mathematician and astronomer. Although Pitiscus worked much in the theological field, his proper abilities concerned mathematics, and particularly trigonometry.

The word "trigonometry" was coined by Pitiscus and first occurs in the title of his work *Trigonometria*. It was first published in Heidelberg in 1595. In 1600, a revised version of his book was published in Augsburg. This book has three sections.

The first section, divided into five books, covers plane and spherical trigonometry.

In the first book he introduced the main definitions and theorems of plane and spherical trigonometry.

In the second book he defined the six trigonometric functions, gave results concerning which properties of a triangle must be known in order to solve it using these trigonometric functions, and also gave techniques to construct tables of the functions. For example he shows how to construct sine tables based on a knowledge of the values of sin 45°, sin 30°, and sin 18°.

The third of the five books is devoted to plane trigonometry, and it consists of six fundamental theorems given with proofs.

The fourth book consists of four fundamental theorems on spherical trigonometry.

The fifth book proves a number of propositions on the trigonometric functions.

The second section consists of tables for all six trigonometric functions. This section has the title *Canon of triangles: tables of sines, tangents, and secants, the radius assumed to be 100000*. The tables give the values to five or six decimal places.

The third section of the work contains ten books discussing: *problems of geodesy, measuring of heights, geography, goniometry, and astronomy.*

Bartholomeo Pitiscus (Ref 77) is sometimes credited with inventing the decimal point, the symbol separating integers from decimal fractions. It was used in his trigonometrical tables and was accepted by John Napier in his logarithmic papers published in 1614 and 1619.

CHAPTER 10

MATHEMATICS DURING THE SCIENTIFIC REVOLUTION (17TH AND 19TH CENTURIES)

The scientific revolution (Ref 78) is a description of the emergence of modern science when developments in mathematics, physics, astronomy, biology, and chemistry transformed the views of society about nature. The scientific revolution took place in Europe towards the end of the Renaissance period and continued through the late 18th century. While the dates are debated, the publication in 1543 *On the Revolutions of the Heavenly Spheres* by Nicolaus Copernicus, is often cited as marking the beginning of the scientific revolution.

In the wake of the Renaissance, the 17th Century saw an unprecedented explosion of mathematical and scientific ideas across Europe, a period sometimes called the Age of Reason. Scientists like Galileo Galilei, Tycho Brahe and Johannes Kepler were making equally revolutionary discoveries in the exploration of the solar system, leading to Kepler's formulation of mathematical laws of planetary motion.

Most of the late 17th century and a good part of the early 18th were taken up by the work of disciples of Newton and Leibniz, who applied their ideas on calculus to solving a variety of problems in physics, astronomy and engineering.

10.1 John Napier (1550–1617)

John Napier (Ref 79) was a Scottish mathematician, physicist, and astronomer. He is best known as the discoverer of logarithms.

John Napier was privately tutored and did not have formal education until he was 13 years old. In 1563, he was sent to St Salvator's College in St. Andrew, Scotland. John did not stay in college very long. He dropped out of school and travelled in mainland Europe to continue his studies. Little is known about those years.

In 1571, Napier returned to Scotland. Many mathematicians at the time were aware of the issues of computation. In particular, John Napier was famous for his devices to assist with computations. He invented so-called "Napier's bones," a manually operated calculating device for calculation of products and quotients of numbers.

The need for complex calculations in the 16th century grew rapidly. A significant part of the difficulties was connected with the multiplication and division of many-valued numbers.

In the course of trigonometric calculations, Napier came up with the idea: to replace laborious multiplication by simple addition, comparing with the help of special tables the geometric and arithmetic progressions, while the geometric will be the original one. Then the division is automatically replaced by an immeasurably simpler and more reliable subtraction.

In 1614, Napier published an essay entitled "Description of the Amazing Table of Logarithms." He described the properties of the logarithms and the seven-digit tables of the logarithms of sines, cosines, and tangents for angles from 0° to 90°, in increments of 1 minute.

The term Napierian logarithm is often used to mean the natural logarithm[1])

$$\text{NapLog}(x) = -10^7 \ln(x/10^7)$$

The Napierian logarithm satisfies identities quite similar to the modern logarithm, such as:

$$\text{NapLog}(xy/10^7) = \text{NapLog}(x) + \text{NapLog}(y)$$

---

[1] The natural logarithm of a number is its logarithm to the base of the mathematical constant e that is an irrational and transcendental number approximately equal to 2.718281828. Napier did not introduce this logarithmic function. However, if it is taken to mean the "logarithms" as originally produced by Napier:

Napier's invention of logarithms was quickly taken up at Gresham College[2] and famous English mathematician Henry Briggs[3] visited Napier in 1615.

The logarithm is the inverse operation to exponentiation. That means the logarithm of a number is the exponent to which another fixed number, the base, must be raised to produce that number. In simple cases the logarithm counts factors in multiplication.

For example, the base 10 logarithm of 1000 is 3, as 10 to the power 3 is 1000 ($1000 = 10 \times 10 \times 10 = 10^3$). 10 is used as a factor three times.

More generally, exponentiation allows any positive real number (a value that represents a quantity along the line) to be raised to any real power, always producing a positive result, so the logarithm can be calculated for any two positive real numbers b and x where b is not equal to 1.

The logarithm of x to base b, denoted $\log_b(x)$, is the unique real number y such that $b^y = x$. For example, $\log_2(64) = 6$, as $64 = 2 \times 2 \times 2 \times 2 \times 2 \times 2 = 2^6$.

The logarithm to base 10, notation is $\log_{10}(x)$, or $\log(x)$, is called the common logarithm and has many applications in science and engineering.

The logarithm to base e, notation is $\log_e(x)$, or $\ln(x)$. It is widely used in mathematics and physics, because of its simpler derivative.

The binary logarithm uses base 2. Notation is $\log_2(x)$. It is commonly used in computer science.

---

[2] Gresham College is an institution of higher learning located in London, England. It does not enroll students and does not award any degrees. It was founded in 1597 under the will of Sir Thomas Gresham (1519–1579), a merchant and financier.

[3] Henry Briggs (1561–1630) is notable for changing the original logarithms, invented by John Napier, into common logarithms with base 10. Common logarithms are also known as Briggsian logarithms, in his honor.

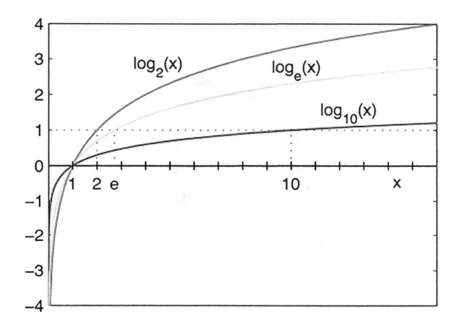

Napier made a common use of the decimal point in mathematics. Decimal point is a symbol used to separate the integer part from the fractional part of a number written in decimal form (numeral system has 10 as its base in every positional numerical system. It is the numerical base most widely used by modern civilizations).

10.2 René Descartes (1596–1650)

René Descartes, in Latin Renatus Cartesius, (Ref 80) was a French philosopher, mathematician, and scientist[4].

In 1607, René entered the Jesuit College, where he was introduced to mathematics and physics, including Galileo's work. After graduation in 1614, he studied for two years at the University of Poitiers[5], earning a Baccalaureate degree and License in civil law (according to his father's wishes that he should become a lawyer).

In 1618, Descartes decided to become a professional military officer and joined the Protestant Dutch States Army (it was the army of the Dutch Republic that existed from 1581 to 1795) and undertook a formal study of military engineering. There he received much encouragement to advance his knowledge of mathematics. Descartes worked on free fall, catenary[6], conic section[7], and fluid statics.

---

[4] A scientist is a person engaging in a systematic activity to acquire knowledge that describes and predicts the natural world.
[5] This university in France, is one of the oldest universities of Europe.
[6] A catenary is the curve that an idealized hanging chain or cable assumes under its own weight when supported only at its ends.
[7] A conic section is a curve obtained as the intersection of the surface of a cone with a plane. Details of the conic sections are described earlier, in Section 4.7.

One of Descartes' most enduring legacies was his Cartesian coordinates, which uses algebra to describe geometry. He invented the convention of representing unknowns in equations by x and y, and knowns by a, and b.

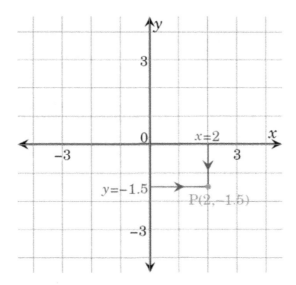

He also pioneered the standard notation that uses superscripts to show the powers or exponents; for example, the 4 used in x^4 to indicate squaring of squaring. He was first to assign a fundamental place for algebra in our system of knowledge and believed that algebra was a method to automate or mechanize reasoning, particularly about abstract, unknown quantities.

European mathematicians had previously viewed geometry as a more fundamental form of mathematics, serving as the foundation of algebra. Algebraic rules were given geometric proofs by mathematicians such as Pacioli, Tartaglia, Ferrari, and Cardano. Equations of degree higher than the third were regarded as unreal, because a three-dimensional form, such as a cube, occupied the largest dimension of reality. Descartes professed that the abstract quantity a^2 could represent length as well as an area. This was in opposition to the teachings of mathematicians, such as Viète, who argued that it could represent only area.

Descartes' work provided the basis for the calculus developed by Isaac Newton and Gottfried Leibniz, who applied infinitesimal calculus to the tangent line problem, thus permitting the evolution of that branch of modern mathematics. His rule of signs is also a commonly used method to determine the number of positive and negative roots of a polynomial.

Descartes' solution of $x^2 + ax = b^2$

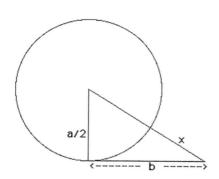

We find x such that $x^2 + ax = b^2$.

First draw a line of length b

Next draw a line of length a/2 perpendicular to this first line at one end of it.

Draw a circle of radius a/2 with centre at the end of this line.

Now join the centre of the circle to the end of the original line.

The distance from the point where it cuts the circle to the end of the line is x

According to the Pythagorean Theorem, in right triangle the square of the hypotenuse (x) is equal to the sum of the squares of the legs ($a/2$) and (b):

$$x^2 = (a/2)^2 + b^2 = a^2/4 + b^2 \text{ and } x = \sqrt{(a^2/4) + b^2}$$

Let's $a = 6$ and $b = 4$, then $x = \sqrt{9 + 16} = \sqrt{25} = 5$

René Descartes is regarded as the "Father of Modern Philosophy" for defining a starting point for existence, "I think, therefore I am" (in Latin: "Cogito, ergo sum").

On February 1, 1650, Descartes contracted pneumonia and died on February 11. He was 53 years old.

10.3 Pierre de Fermat (1601–1665)

**Pierre de Fermat
(1601-1665)**

Pierre de Fermat (Ref 81) was a French mathematician who is given credit for early developments that led to infinitesimal calculus, including his technique of adequality. In particular, he is recognized for his discovery of an original method of finding the greatest and the smallest ordinates of curved lines, which is analogous to that of differential calculus, then unknown, and his research into number theory. He made notable contributions to analytic geometry, probability, and optics. He is best known for his Fermat's Last Theorem in number theory.

There is little evidence concerning his school education, but it was probably at the Collège de Navarre in Montauban. He attended the University of Orléans in 1623 and received bachelor's in civil law in 1626.

Fermat was the first person known to have evaluated the integral of general power functions[8]. With his method, he was able to reduce this evaluation to the sum of geometric series[9]. The resulting formula was helpful to Newton, and then Leibniz, when they independently developed the fundamental theorem of calculus.

Historically, geometric series played an important role in the early development of calculus, and they continue to be central in the study of convergence of series. Geometric series are used throughout mathematics, and they have important applications in physics, computer science, engineering, queueing theory[10], biology, economics, and finance.

In 1637, Fermat conjectured[11] his Last Theorem, which states that no three positive integers a, b, and c satisfy the equation: $a^n + b^n = cn$ for any integer value of n greater than 2.

---

[8] In mathematics, a power function is a function of the form $f(x) = x^p$ where p is constant and x is a variable. In general, p can belong to one of several classes of numbers, such as the positive and negative integers. They are a fundamental concept in algebra and pre-calculus, leading up to the formation of polynomials. Their general form is $f(x) = cx^p$, where c is also a constant.

[9] In mathematics, a geometric series is a series with a constant ratio between successive terms. For example, the series $1/2 + 1/4 + 1/8 + ...$ is geometric, because each successive term can be obtained by multiplying the previous term by $1/2$.

[10] Queueing theory is the mathematical study of waiting lines, or queues. Queueing theory is generally considered a branch of operations research because the results are often used when making business decisions about the resources needed to provide a service.

[11] In mathematics, a conjecture is a conclusion or proposition based on incomplete information, for which no proof has been found.

The first successful proof of Last Theorem was released in 1994 by Andrew Wiles[12]), and formally published in 1995, after 358 years of effort by mathematicians.

In 1640, Fermat published the less famous, but much more fundamental, Little Theorem. This theorem states that if p is a prime number, then for any integer a, the number ($a^p - a$) is an integer multiple of p.

For example, if $a = 2$ and $p = 7$, then $2^7 = 128$, and $128 - 2 = 126 = 7 \times 18$ is an integer multiple of 7.

Pierre Fermat, like many mathematicians before and after him, was occupied by the problem of prime numbers. He searched for expressions generating prime numbers and stated that such a generator is the formula:

$$F_n = 2^{2^n} + 1,$$

where n is not negative integer.
The first six Fermat numbers are:

$n = 0$, then $2^0 = 1$, and the first number $F_0 = 2^1 + 1 = 3$
$n = 1$, then $2^1 = 2$, and the second number $F_1 = 2^2 + 1 = 5$
$n = 2$, then $2^2 = 4$, and the third number $F_2 = 2^4 + 1 = 17$
$n = 3$, then $2^3 = 8$, and the fourth number $F_3 = 2^8 + 1 = 257$
$n = 4$, then $2^4 = 16$, and the fifth number $F_4 = 2^{16} + 1 = 65{,}537$
$n = 5$, then $2^5 = 32$, and the sixth number $F_5 = 2^{32} + 1 = 4{,}294{,}967{,}297$

Indeed, for $n = 0, 1, 2, 3,$ and 4 we obtain the prime numbers 3, 5, 17, 257, and 65,537. Pierre de Fermat checked this and concluded that his formula generates prime numbers.

In 1739, Leonhard Euler showed that Fermat number F_5 is not a prime number, because it is divided by 641:

$4{,}294{,}967{,}297/641 = 6{,}700{,}417$ (67,700,417 is a prime number).

---

[12] Sir Andrew John Wiles (born April 11, 1953) is a British mathematician and a Royal Society Professor at the University of Oxford, specializing in number theory. He received the 2016 Abel Prize[13]) for proving Fermat's Last Theorem.

[13] In brief, about the Abel Prize and its laureates, you can read in Section 11.4.

Contemporaries characterized Fermat as an honest, neat, balanced, and affable person, brilliantly erudite both in mathematics and in the humanities, an expert in many ancient and living languages, in which he wrote good poetry. He died on January 12, 1665. He was 63 years old.

There is a curious episode (Ref 82, p. 107) translated from Russian into English by the author of this book. The boy, who performed the instant account session, was asked whether the number 4,294,967,297 is a prime. After a short pause, he replied, "This number is not a prime, because it is divided by 641". He couldn't explain how <u>he came to this conclusion. The name of that boy is Zerah Colburn[14])."</u>

Zerah is considered as a child prodigy of the 19th century who gained fame as a mental calculator. He was thought to be intellectually disabled until the age of six. However, after six weeks of schooling his father overheard him repeating aloud the multiplication tables.

His father wasn't sure whether or not he learned the tables from his older brothers and sisters, but he decided to test him further on the mathematical abilities.

His father discovered that there was something special about Zerah. He asked Zerah to multiply 13 by 97, and his son answered in a few seconds, "1261."

When he was seven years old, he took six seconds to give the numbers of hours in thirty-eight years, two months, and seven days.

His father capitalized on his boy's talents by taking Zerah around the country and eventually abroad, demonstrating the boy's exceptional abilities. Despite regular classes at school, he showed talent in languages.

In 1835, at the age of 31, he was appointed professor of languages at Dartmouth College in Hanover, New Hampshire. In 1833, he published his autobiography. He died on March 2, 1839, of tuberculosis at the age of 34.

---

[14] Zerah Colburn (Ref 83) was born on September 1, 1804, in the small town of Cabot, Vermont, USA.

10.4 Blaise Pascal (1623–1662)

Blaise Pascal (Ref 84) was a French mathematician, physicist, inventor, writer, and philosopher. He was a child prodigy who was educated by his father. Pascal's earliest work was in the natural and applied sciences where he made important contributions to the study of fluids, and clarified the concepts of pressure and vacuum.

In 1642, while still a teenager, he started some pioneering work on calculating machines. After three years of effort he developed and built 20 machines (called Pascal's calculators). This is one of them.

Pascal was the second inventor of the mechanical calculators. The first inventor of this kind of calculators was Wilhelm Schickard (1592–1635), a German professor of Hebrew and Astronomy.

In 1623 and 1624, Wilhelm Schickard sent two letters to a German astronomer Johannes Kepler (1571–1630) about his design and construction of what he referred to as an "arithmetical instrument" that he has invented. The instrument was designed to assist in four basic functions of arithmetic: addition, subtraction, multiplication and division. Amongst its uses, Schickard suggested that it would help in the laborious task of calculating astronomical tables. The calculator could add and subtract six-digit numbers and indicated an overflow of this capacity by ringing a bell.

Pascal practiced mathematics all his life. In 1653, Pascal wrote "Treatise on the Arithmetical Triangle" in which he described a convenient tabular presentation for binomial coefficients, now called Pascal's triangle.

The following example (Ref 85, pp. 90, 91) shows the first six rows of Pascal's triangle:

$$\begin{array}{ccccccc}
& & & 1 & & & \\
& & 1 & & 1 & & \\
& & 1 & 2 & 1 & & \\
& 1 & 3 & & 3 & 1 & \\
& 1 & 4 & 6 & 4 & 1 & \\
1 & 5 & 10 & & 10 & 5 & 1 \\
\end{array}$$

On the first row is only the number **1**. Then, to construct the elements of the following rows, add the number directly above and to the left (if any) and the number directly above and to the right (if any) to find the new value. For example, the numbers **1** and **2** in the third row are added to produce **3** in the fourth row, and so on. The coefficient **2** and higher are shown in the following equations:

$$(x + y)^2 = x^2 + 2xy + y^2$$
$$(x + y)^3 = x^3 + 3x^2y + 3xy^2 + y^3$$
$$(x + y)^4 = x^4 + 4x^3y + 6x^2y^2 + 4xy^3 + y^4$$
$$(x + y)^5 = x^5 + 5x^4y + 10x^3y^2 + 10x^2y^3 + 5xy^4 + y^5$$

Pascal in his letter to Pierre de Fermat, described his method of solving gambling problems and asked Fermat if he agreed with such method. Fermat and Pascal investigated gambling problems using certain proportions, which later were defined as probabilities. The mathematical theory of probabilities was born from that collaboration.

Pascal made significant contribution in studying fluids and laid the foundation for hydrostatics and hydrodynamics. Calculation of the fluid pressure is based on physical law called Pascal's Principle. It states that the pressure exerted by a fluid at a given depth is *equal in all directions*. According to such physical law, Pascal invented hydraulic press. No surprises that modern hydraulic presses utilize the same principle:

Force (N) acting on a plunger is a product of the fluid pressure (P) generated by a pump, and a cross-sectional area (A) of the plunger, N = P × A (Ref 85, p.90).

On October 8, 1971, the 14th General Conference on Weights and Measure adopted definition of the **pascal** as a unit of pressure and stress in the metric system of measurement (Ref 85, pp. 85–93). Mathematical expression of the **pascal** (Pa) is:

$$Pa = N/m^2$$

where N (newton) is a unit of force,
m (meter) is a unit of length.

Meter is a base unit, and newton is a derived unit of the metric system. About the metric system can be found in the book "International System of Units (IS)". This book was published in 2012 by Industrial Press, Inc. The author of this book is Edmund Isakov.

In 1972, a Swiss computer scientist Niklaus Wirth (born on February 15, 1934) named his new computer language "Pascal" and insisted that it to be spelled Pascal, not PASCAL (Ref 85, p.92).

10.5 Gottfried Wilhelm Leibniz (1646–1716)

Gottfried Wilhelm Leibniz (Ref 86) was a German polymath who occupies a prominent place in the history of philosophy and contribution to mathematics, especially, in developing differential and integral calculus independently of Isaac Newton.

In 1679, Leibniz devised the binary number system, which later he described in his article *Explanation of Binary Arithmetic*. All calculations were carried out with only two digits 0 and 1.

Systems related to binary numbers, have appeared earlier in multiple cultures including ancient Egypt, China, and India. Leibniz was specifically inspired by the *I Ching*, an ancient Chinese text. The *I Ching* dates from the 9th century BC in China. The binary notation in *I Ching* is used to interpret its quaternary technique. Quaternary is the base-4 numeral system. It uses the digits 0, 1, 2, 3 to represent any real number.

Leibniz's system uses digits 0 and 1. An example of Leibniz's binary numeral system is as follows:

0 0 0 1   numerical value 2^0
0 0 1 0   numerical value 2^1
0 1 0 0   numerical value 2^2
1 0 0 0   numerical value 2^3

The binary, or base-2, number system expresses all number as a combination of the digits 0 and 1.

Comparison between decimal and binary systems:

Decimal	1	2	3	4	5	6	7	8	9	10
Binary	1	10	11	100	101	110	111	1000	1001	1010

The base-2 system is a positional notation with a radix of 2. Because of its straightforward implementation in digital electronic circuitry using logic gates, the binary system is used internally by almost all modern computers and computer-based devices. Each digit is referred to as a bit.

Leibniz became one of the most prolific inventors in the field of mechanical calculators. While working on adding automatic multiplication and division to Pascal's calculator, he was the first to describe a pinwheel calculator in 1685 and invented the Leibniz wheel, used in the arithmometer, the first mass-produced mechanical calculator.

Between his work (Ref 87) on philosophy, logic, and his day job as a politician and representative of the royal house of Hanover, Leibniz still found time to work on mathematics. He was perhaps the first to explicitly employ the mathematical notion of a function to denote geometric concepts derived from a curve, and he developed a system of infinitesimal calculus, independently of his contemporary Sir Isaac Newton.

Leibniz was a member of the Royal Society in London, and aware of Newton's work on calculus. During the 1670s (slightly later than Newton's early work), Leibniz developed a very similar theory of calculus, apparently completely independently. Within the short period of about two months he had developed a complete theory of differential calculus and integral calculus.

Unlike Newton, Leibniz was more than happy to publish his work, and so Europe first heard about calculus from Leibniz in 1684, and not from Newton (who published nothing on the subject until 1693). When the Royal Society was asked to adjudicate between the rival claims of the two men over the development of the theory of calculus, they gave credit for the first discovery to Newton, and credit for the first publication to Leibniz. However, the Royal Society, by then under the rather biased presidency of Newton himself, later also accused Leibniz of plagiarism, a slur from which Leibniz never really recovered.

When differentiating, Leibniz used dx and dy to indicate infinitesimal increments in the independent and dependent variables: dy/dx or $d(f(x))/dx$

Currently, we use Leibniz's notation, which contains an extended "S" to indicate the sum of infinitely many infinitesimal quantities $f(x)$ for each infinitesimal increment dx, between the two stated limits a and b:

$$\int_a^b f(x)dx$$

Leibniz rediscovered a method of arranging linear equations[1] into an array, now called a matrix[2]. A similar method had been developed by Chinese mathematicians <u>almost two millennia earlier but had long fallen into disuse.</u>

1) Linear equation is an algebraic equation in which each term is either a constant or the product of a constant and a single variable. System of

linear equations is a collection of two or more linear equations involving the same set of variables. For example:

$$3x + 2y - z = 1$$
$$2x - 2y + 4z = -2$$
$$-x + \frac{1}{2}y - z = 0$$

2) In mathematics, a matrix (plural matrices) is a rectangular array of numbers, symbols, or expressions, arranged in rows and columns. For example, the dimensions of the matrix below are 2 × 3 (read "two by three"), because there are two rows and three columns:

$$\begin{matrix} 1 & 3 & 4 \\ 5 & 7 & 8 \end{matrix}$$

Leibniz is considered the most important logician among Aristotle (384 –322 BC) in Ancient Greece; George Boole (1815–1864), an English mathematician, philosopher, logician; and Augustus De Morgan (1806–1871), a British mathematician and logician.

10.6 Isaac Newton (1643–1727)

Sir Isaac Newton (Ref 88) was an English mathematician, astronomer, and physicist (described in his own day as a "natural philosopher") who is widely recognized as one of the most influential scientists of all time and a key figure in the scientific revolution. Newton laid the foundations of classical mechanics and also made seminal contributions to optics. He shares credit with Gottfried Wilhelm Leibniz for developing the infinitesimal calculus.

During the time of the Great Plague of 1665-6, Newton (Ref 89) developed a new theory of light, discovered gravitation, and pioneered a revolutionary new approach to mathematics: infinitesimal calculus. His calculus allowed mathematicians and engineers to make sense of the motion and dynamic change in the changing world around us, such as the orbits of planets, the motion of fluids, etc.

Newton (and his contemporary Gottfried Leibniz independently) calculated a derivative function $f'(x)$ which gives the slope at any point of a function $f(x)$. This process of calculating the slope or derivative of a curve or function is called differential calculus or differentiation (or, in Newton's terminology, the "method of fluxions"—he called the instantaneous rate of change at a particular point on a curve the "fluxion," and the changing values of x and y the "fluents"). For instance, the derivative of a straight line of the type $f(x) = 4x$ is just 4; the derivative of a squared function $f(x) = x^2$ is $2x$; the derivative of cubic function $f(x) = x^3$ is $3x^2$, etc. Generalizing, the derivative of any power function $f(x) = x^n$ is nx^{n-1}. Other derivative functions can be stated, according to certain rules, for exponential and logarithmic functions, trigonometric functions such as sin (x), cos (x), etc. So that a derivative function can be stated for any curve without discontinuities. For example, the derivative of the curve $f(x) = x^4 - 5x^3 + \sin(x^2)$ would be:

$$f'(x) = 4x^3 - 15x^2 + 2x \cos(x^2)$$

Leibniz's and Newton's notation for Calculus (Ref 87):

Calculus Notation

Differentiation

Newton used a dot over the dependent variable to indicate a derivative, two dots for a second derivative, etc.	$\dot{y} = \dfrac{dy}{dt}$
Leibniz just used dx and dy to indicate infinitesimal increments in the independent and dependent variables.	$\dfrac{d(f(x))}{dx}$ or $\dfrac{dy}{dx}$
The use of the prime mark for derivatives dates from Joseph Louis Lagrange's work on differential calculus in the late 18th / early 19th Century.	$f'\ f''$

Integration

Newton never used one consistent notation for integration, sometimes using a bar above a variable and sometimes putting the variable in a box.	$\overline{x}$ or $\boxed{x}$
We have come to use Leibniz's notation, which uses an extended 'S' to indicate the sum of infinitely many infinitesimal quantities $f(x)$ for each infinitesimal increment dx, between the two stated limits a and b.	$\displaystyle\int_a^b f(x)\,dx$

Newton is considered by many to be one of the most influential men in human history. In 1687, Newton published his "Principia" or "The Mathematical Principles of Natural Philosophy," generally recognized as the greatest scientific book ever written. In it, he presented his theories of motion, gravity and mechanics, explained the eccentric orbits of comets, the tides and their variations, the precession of the Earth's axis and the motion of the Moon.

Later in life, he wrote a number of religious tracts dealing with the literal interpretation of the Bible, devoted a great deal of time to alchemy, acted as Member of Parliament for some years, and became perhaps the best-known Master of the Royal Mint in 1699, a position he held until his death in 1727.

In 1703, Newton became president of the Royal Society and, in 1705, became the first scientist ever to be knighted. Mercury poisoning from his alchemical pursuits perhaps explained Newton's eccentricity in later life, and possibly also his eventual death.

One day, Newton said, "If I have seen further than others, it is by standing upon the shoulders of giants."

On October 21, 1948, the 9th General Conference on Weights and Measure adopted a definition of the newton as a unit of the force in the metric system of measurement: *A force of one newton accelerates a mass of one kilogram at the rate of one meter per second per second* (Ref 85, pp.74–84).

Mathematical expression of the newton (N) is:

$$N = kg \times m/s^2$$

where kg (kilogram) is a unit of mass,
m (meter) is a unit of length,
s (second) is a unit of time

Kilogram, meter, and second are the base units; and *newton* is a derived unit of force in the International System of Units (SI).

More information on the life and scientific work of Isaac Newton is provided in Appendix 9.

10.7 Leonhard Euler (1707–1783)

Portrait of Leonhard Euler by Jakob Emanuel Handmann

Leonhard Euler (Ref 90) was a Swiss mathematician, physicist, astronomer, and logician. Historically, logic has been studied in philosophy (since ancient times) and mathematics (since the mid-1800s), and recently logic has been studied in computer science, linguistics, psychology, and other fields.

Leonhard Euler was born on April 15, 1707, in Basel, Switzerland, where he obtained a formal education. In 1720, aged thirteen, he enrolled at the University of Basel. In 1723, he received a master's degree in philosophy, defending his thesis, in which he compared Descartes's philosophy with Newton's philosophy. During that time, Leonhard attended Saturday afternoon lessons from Johann

Bernoulli[15], who <u>quickly discovered that his new pupil had incredible talent for mathematics.</u>

At that time, Leonhard's main studies included theology, Greek, and Hebrew at his father's demand because he wanted his son to become a pastor. Johann Bernoulli convinced Leonhard's father that his son was destined to become a great mathematician.

In 1726, Leonhard Euler completed a dissertation on the propagation of sound. At that time, he was unsuccessfully attempting to obtain a position at the University of Basel.

Around that time, Leonhard Euler's friends Daniel and Nicolaus Bernoulli were working at the Imperial Russian Academy of Sciences in Saint Petersburg. In 1726, Daniel assumed position in the mathematics/physics division and recommended that his post in physiology that he had vacated be filled by his friend Leonhard Euler. In November 1726, nineteen-year-old Euler eagerly accepted the offer.

Euler arrived in Saint Petersburg in May 1727. He was promoted from his junior post in the medical department of the academy to a position in the mathematics department. He lodged with Daniel Bernoulli with whom he often worked in close collaboration. Euler mastered Russian and settled into life in Saint Petersburg. He also took on an additional job as a medic in the Russian Navy.

Euler swiftly rose through the ranks in the academy and was appointed professor of physics in 1731. Two years later, Daniel Bernoulli, who was fed up with the censorship and hostility he faced at Saint Petersburg, left for Basel. Euler succeeded him as the head of the mathematics department.

Concerned about the continuing turmoil in Russia, Euler left Saint Petersburg in June 1741 to take up a post at the Berlin Academy, which he had been offered by Frederick the Great of Prussia (Frederick II, King of Prussia from 1740 until 1786). Euler lived for twenty-five years in Berlin, where he wrote over 380 articles. There Euler published the

---

[15] Johann Bernoulli (1667–1748) was a Swiss mathematician and one of the many prominent mathematicians in the Bernoulli family. He is known for his contributions to infinitesimal calculus and educating Leonhard Euler in the pupil's youth.

two works for which he would become most renowned: *Introduction to the analysis of the infinite* (written in Latin and published in 1748), which lays the foundations of mathematical functions, and *Foundations of differential calculus* (written in Latin in 1748 and published in 1755) that lays the groundwork for the differential calculus.

In 1755, Leonhard Euler was elected a foreign member of the Royal Swedish Academy of Sciences.

The political situation in Russia stabilized after Catherine II (also known as Catherine the Great) was coroneted Empress of Russia. In 1766, Euler accepted the invitation from Imperial Russian Academy of Sciences and returned to Saint Petersburg. His conditions were quite enormous: 3000 rubles annual salary, a pension for his wife, and the promise of high-ranking appointments for his sons. All of these requests were granted. He spent the rest of his life in Russia.

On September 18, 1783, after a lunch with his family, Euler was discussing the newly discovered planet Uranus and its orbit with a fellow academician when he collapsed from a brain hemorrhage. Leonhard Euler died a few hours later. He was 76 years old.

Euler introduced and popularized several notational conventions through his numerous and widely circulated textbooks. Especially, he introduced the concept of a function (relation between a set of inputs and a set of permissible outputs with the property that each input is related to exactly one output). Euler was the first to write $f(x)$ to denote the function f applied to the argument x.

Euler started to use the letter e for the constant in 1727 or 1728, in an unpublished paper on explosive forces in cannons. The first appearance of e in a publication was in Euler's *Mechanica* in 1736. The value of e, he calculated by the following formula, and used that value as the base of the natural logarithm:

$$e = \sum_{n=0}^{\infty} \frac{1}{n!} = \frac{1}{1} + \frac{1}{1} + \frac{1}{1 \cdot 2} + \frac{1}{1 \cdot 2 \cdot 3} + \cdots$$

Euler calculated its value to 23 decimal places:

$e = 2.71828182845904523536028$

The number e is called *Euler's number*, and like π is a transcendental number. The letters e and π are the most important numbers in mathematics.

There are many ways of calculating the value of e, but none of them give an exact answer, because number e is irrational (not the ratio of two integers). For example, the value of $(1 + 1/n)^n$ approaches e as n gets bigger and bigger:

n	$(1 + 1/n)^n$
10	2.593742460
100	2.704813829
1,000	2.716923932
10,000	2.718145936
100,000	2.718268303
1,000,000	2.718281378

Euler used Greek letter Σ for summation, and the letter i to denote the imaginary unit, equal to the square root of −1 ($i = \sqrt{-1}$). The use of the Greek letter π to denote the ratio of the circumference of a circle to its diameter was also popularized by Euler. However, the first person who suggested using the letter π was a Welsh mathematician William Jones (1675–1749). Euler also introduced the modern notation for the trigonometric functions: sine (*sin*), cosine (*cos*), tangent (*tan*), cotangent (*cot*), secant (*sec*), and cosecant (*csc*).

Euler defined the exponential functions for complex numbers. Complex number is a number that can be expressed in the form $a + bi$, where a and b are real numbers, and i is the imaginary unit, which satisfies the equation $i^2 = -1$: a is the real part of the complex number, and b is the imaginary part. He also discovered relation between the exponential functions and the trigonometric functions. For any real number φ (taken to be radians), Euler's formula states that the complex exponential function satisfies:

$$e^{i\varphi} = \cos \varphi + i \sin \varphi$$

A special case of the above formula is Euler's identity or Euler's equation, which is the equality:

$$e^{i\pi} + 1 = 0$$

Richard P. Feynman (1918–1988), an American theoretical physicist, called the above equality the most remarkable formula in mathematics for its single uses of the notions of addition, multiplication, exponentiation, and equality; and the single uses of the important constants 0, 1, e, i, and π.

In 1988, readers of the *Mathematical Intelligencer* (a mathematical journal that aims at a conversational and scholarly tone) voted it "the Most Beautiful Mathematical Formula Ever." In that poll, among chosen the top five formulas, three of them were developed by Euler.

Euler is well known in dealing with limits, differentiations, integration, infinite series, and analytic functions for his frequent use and development of power series, the expression of functions as sums of infinitely many terms, such as:

$$e^x = \sum_{n=0}^{\infty} \frac{x^n}{n!} = \lim_{n \to \infty} \left(\frac{1}{0!} + \frac{x}{1!} + \frac{x^2}{2!} + \cdots + \frac{x^n}{n!} \right).$$

Euler significantly advanced the theory of series. The mathematical world was greatly impressed by the series that were first summed up by him, including a series of inverse squares. No one had been solved such series before Euler. For example, this series:

$$\sum_{n=1}^{\infty} \frac{1}{n^2} = \lim_{n \to \infty} \left(\frac{1}{1^2} + \frac{1}{2^2} + \frac{1}{3^2} + \cdots + \frac{1}{n^2} \right) = \frac{\pi^2}{6}.$$

Let's count the right-hand side of this series:

$\pi^2/6 = 9.869604/6 = 1.644934$

This means that no matter how much we sum up the terms of this series, we never get the number 1.644934.

Let's calculate the sum of the three terms of the above series:

$1/1^2 + 1/2^2 + 1/3^2 = 1 + 1/4 + 1/9 = 1 + 0.25 + 0.111111 = 1.361111$

Let's add the three more terms to the left-hand side of this formula:

$1/4^2 + 1/5^2 + 1/6^2 = 1/16 + 1/25 + 1/36 = 0.0625 + 0.004 + 0.0278 = 0.0943$

There is the sum of these six terms:

$1.361111 + 0.0943 = 1.455411$

Let's calculate the percentage (x) of the sum of six terms (1.455411) compared to the number 1.644934:

$x = 1.455411 \times 100 / 1.644934 = 88.5\%$

Dear readers, if you are interested in continuing this process (adding terms to the left-hand part of this series of inverse squares and comparing the summation with the right-hand part) please, do it.

Euler made important discovery in many branches of mathematics: infinitesimal calculus, graph theory, topology, and analytic number theory. He also introduced much of the modern mathematical terminology and notation for mathematical analysis. He was one of the most eminent mathematicians of the 18th century.

Euler is also known for his work on mechanics, astronomy, weights and measures, fluid dynamics, cartography, optics, and even the theory of music.

In 1755, Euler was elected a Foreign Member of the Royal Swedish Academy of Sciences. In 1782, Euler was elected a Foreign Honorary Member of the American Academy of Arts and Sciences. He is widely considered to be the most prolific mathematician of all widely time.

The complete collection of Euler's works was publishing since 1909 by the Swiss Society of Naturalists and still is incomplete. It was planned to publish 75 volumes. As of today, the Society published 73 volumes (Ref 91):

- 29 volumes on mathematics
- 31 volumes on mechanics and astronomy
- 13 volumes on physics.

10.8 Joseph-Louis Lagrange (1736–1813)

Portrait of Joseph-Louis Lagrange by Robert Hart (British Museum)

Joseph-Louis Lagrange (Ref 92) was born in Turin, Italy, and baptized in the name of Giuseppe Ludovico Lagrangia. Lagrange's family had French connections on his father's side, his great-grandfather being a French cavalry captain who left France to work for the Duke of Savoy. Lagrange always leant towards his French ancestry. Being young, he used to sign himself as Lodovico LaGrange or Luigi Lagrange, using the French form of his family name. Lagrange is usually considered to be a French mathematician, but the Italian Encyclopedia refers to him as an Italian mathematician.

Lagrange (Ref 93) studied at the University of Turin, and his favorite subject was classical Latin. At first, he had no great enthusiasm for mathematics, finding Greek geometry rather dull. At the age of seventeen, he showed the taste for mathematics, when he read a paper on the use of algebra in optics by Edmond Halley[16]).

Alone and unaided Lagrange threw himself into mathematical studies, and about a year of hard work he was already an accomplished mathematician.

In 1755, Charles Emmanuel III[17] appointed Lagrange to serve as the mathematics assistant professor at the Royal Military Academy in Turin, where he taught courses in calculus and mechanics to support the ballistics theories of Leonhard Euler and Benjamin Robins (1707–1751), an English mathematician and military engineer. In that capacity, Lagrange was the first to teach calculus in an engineering school.

Lagrange is one of the founders of the calculus of variations (it is a type of mathematics involving maxima and minima). Starting in 1754, he worked on the problem of isochrone curve, for which the time, taken by an object sliding without friction in uniform gravity to its lowest point, is independent of its starting point. The curve of this kind is called a cycloid (it traced by a point on the rim of a circular wheel as the wheel rolls along a straight line without slipping).

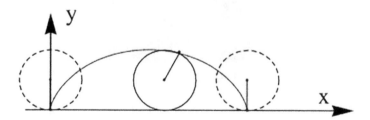

Lagrange discovered a method of maximizing and minimizing functional in a way similar to finding extrema of functions. He wrote several letters to Leonhard Euler between 1754 and 1756 describing his

---

[16] Edmond Halley (1656–1744) was an English astronomer who calculated the orbit of the comet, which later was named after him.

[17] Charles Emmanuel III (1701–1773) was the Duke of Savoy and King of Sardinia.

results, leading to the Euler equations of variational calculus. However, the results of Lagrange proved to be simpler than that of Euler. He was very impressed with Lagrange's results.

In 1756, Euler and Maupertuis[18], seeing Lagrange's mathematical talent, tried to persuade him to come to Berlin, but he had no such intention and refused the offer.

In 1765, d'Alembert[19] interceded on Lagrange's behalf with Frederick II, King of Prussia, and by letter, asked Lagrange to leave Turin for more prestigious position in Berlin. Lagrange again turned down the offer, responding, "*It seems to me that Berlin would not be at all suitable for me while M. Euler is there*".

In 1766, Euler left Berlin for Saint Petersburg, and Frederick II himself wrote to Lagrange expressing the wish of "the greatest king in Europe" to have "the greatest mathematician in Europe" a resident at his court. Lagrange was finally persuaded, and he spent the next twenty years in Prussia, where he had written not only a lot of articles published in the Berlin and Turin transactions, but also his monumental work, the *Analytical mechanics*.

Lagrange was very active scientifically during his twenty years spent in Berlin. Except his remarkable *Analytical mechanics*, he had submitted about two hundred papers to the Academy of Turin, the Berlin Academy, and the French Academy.

Most of Lagrange's papers were on algebra, number theory, analytical geometry, partial differential equations, and the others. Among the works on number theory there is a theorem that he proved in 1770. The theorem states that every positive integer can be represented as the sum of four integer squares. For example:

$$4 = 1^2 + 1^2 + 1^2 + 1^2 = 1 + 1 + 1 + 1$$
$$11 = 3^2 + 1^2 + 1^2 + 0^2 = 9 + 1 + 1 + 0$$
$$51 = 5^2 + 4^2 + 3^2 + 1^2 = 25 + 16 + 9 + 1$$

---

[18] Pierre Louis Maupertuis (1698–1759) was a French mathematician, philosopher and public intellectual.

[19] Jean-Baptiste d'Alembert (1717–1783) was a French mathematician, physicist, philosopher, and music theorist.

$$111 = 10^2 + 3^2 + 1^2 + 1^2 = 100 + 9 + 1 + 1$$
$$230 = 9^2 + 8^2 + 7^2 + 6^2 = 81 + 64 + 49 + 36$$

In 1786, Lagrange accepted the offer of the King of France Louis XVI to move to Paris. He was received with every mark of distinction and special apartments in the Louvre were prepared for his reception, and soon he became a member of the French Academy of Science.

At the beginning of his residence in Paris, he was seized with an attack of melancholy, and even the printed copy of his *Analytical mechanics* on which he had worked for a quarter of a century lay for more than two years unopened on his desk. Curiosity as to the results of the French revolution[20] first stirred him out of his lethargy, a curiosity which soon turned to alarm as the revolution developed.

In France, the period from June 1793 to September 1794, is known as the Reign of Terror. During this period, almost 17 thousand people were killed in the country, including more than 2,600 people in Paris. On May 4, 1794, 27 tax-collecting officials and the famous scientist Lavoisier[21] were arrested, sentenced to death and guillotined on the afternoon after the trial. Lagrange said on the death of Lavoisier, *"It took only a moment to cause this head to fall and hundred years will not suffice to produce such like."*

Lagrange was never in any danger. Different revolutionary government, and at a later time, Napoleon awarded him with honors and distinctions. In the 1790s, Lagrange was offered the presidency of the Commission for the reform of weights and measures. It was largely owing to Lagrange's influence that the final choice of the unit system

---

[20] The French Revolution was a period of far-reaching social and political upheaval in France that lasted from 1789 until 1799.

[21] Antoine Lavoisier (1743–1794) is considered the "father of modern chemistry." He coined the terms *oxygen* and *hydrogen* and had proven that these gases are elements. He discovered that matter may change its form or shape, but its mass always remain the same. When Lavoisier requested time to complete some scientific work, the presiding judge answered, *"The Republic needs neither scientists nor chemists, the course of justice cannot be delayed"* (Ref 85, pp.9, 10, 22, 122, 291).

of *meter* and *kilogram* were settled, and the decimal subdivision was finally accepted by the commission of 1799.

In 1794, Lagrange was appointed professor of the École Polytechnique, a French public institution of higher education and research in a suburb of Paris. He lectured there on the differential calculus. His lectures, described by mathematicians who had to attend them, were almost perfect both in form and matter.

In 1795, Lagrange was appointed to a mathematical chair at École Normale (higher education establishment). His lectures were concrete, interesting and liked by students. The lectures there were published because the professors had to "pledge themselves to representatives of the people and to each other neither to read nor to repeat form memory," and the discourses were ordered to be taken down in shorthand to enable the deputies to see how the professors acquitted themselves.

Lagrange lived and worked in Paris from 1786 until his death in 1813. His work in Paris covered many topics: differential calculus and calculus of variations, infinitesimals, number theory, and celestial mechanics.

In 1798, Lagrange published his work on the number theory *Resolution of numerical equations*. There he gives the method of approximating to the real roots of an equation by means of continued fractions and outlines several other theorems.

Lagrange is best known for his book *Analytical mechanics*, where he transformed Newtonian mechanics (also known as classical mechanics) into a branch of analysis, Lagrangian mechanics (a reformulation of classical mechanics), as it is now called, and presented the mechanical principles as simple results of the variational calculus.

In 1810, Lagrange began revision of the *Analytical mechanics*, but he was able to complete only about two-thirds of it before his death in Paris on April 10, 1813. He was buried in the Panthéon in Paris.

Lagrange name is one of the 72 names of the greatest scientists and engineers of France, placed on the first floor of the Eiffel Tower.

10.9 Pierre-Simon Laplace (1749–1827)

Portrait of Pierre-Simon Laplace by Jean-Baptiste Paulin Guérin (1838)

Pierre-Simon Laplace (Ref 94) was a famous French scholar whose work was important to the development of mathematics, physics, and astronomy. He summarized and extended the work of his predecessors in his five-volume *Celestial Mechanics*. This work translated the geometric study of classical mechanics to one based on calculus, opening up a broader range of problems.

Pierre-Simon Laplace was a born on March 23, 1749, in Beaumont-en-Auge, a village in Normandy region in France. He attended a Benedictine priory school in Beaumont-en-Auge, as a day pupil, between the ages of 7 and 16. His father expected him to make a career in the Church and indeed either the Church or the army were the usual destinations of pupils at the priory school. At the age of 16 Laplace entered Caen University. As he was still intending to enter the Church, he enrolled to study theology. However, during his two years at the University of Caen, Laplace discovered his mathematical talents and his love of the subject.

At the university, Laplace was mentored by two enthusiastic teachers of mathematics who awoke his eagerness on the integral calculus with infinitely small differences and finite differences. Laplace's brilliance as a mathematician was quickly recognized and while still at the University of Caen Laplace wrote a memoir, which provided the first communion between him and Lagrange. Sometime earlier, Lagrange founded a journal *Miscellaneous Taurinensia* and published Laplace's memoir. About this time, Laplace recognized that he had no vocation for the priesthood, he decided to become a professional mathematician.

In 1768, Laplace left for Paris. He took with him a letter of introduction to Jean-Baptiste d'Alembert to Le Canu, his teacher at Caen. At the time, d'Alembert was supreme in scientific circles. Jean-Baptiste d'Alembert received him rather poorly, and to get rid of Laplace gave him a thick mathematics book, saying to come back when he had read it. When Laplace came back a few days later, d'Alembert was even less friendly and did not hide his opinion that it was impossible that Laplace could have read and understood the book. But upon questioning him, d'Alembert realized that it was true, and from that time he took Laplace under his care. Another case when Laplace solved overnight a problem that d'Alembert set him for submission the following week, then solved a harder problem the following night. D'Alembert was impressed and recommended him for a teaching place in the École Militaire (Military School). With a secure income and undemanding teaching, Laplace threw himself into original research and for the next 17 years, 1771–1787, he produced much of original work in astronomy.

In 1771, Laplace published his work on differential equations, finite differences, and was thinking about the mathematical and philosophical concepts of probability and statistics. On March 31, 1773, Laplace was elected associate member of the French Academy of Sciences, a place where he worked on mathematics and astronomy. Before his election to the academy, he had already drafted two papers that established his reputation. The first paper, *Memory on the probability of causes by events* was published in 1774. In the second paper (1776), Laplace began his systematic work on celestial mechanics and the stability of the solar system.

In 1812, Laplace issued his *Analytical theory of probabilities*, in which he laid down many fundamental results in statistics. Laplace's proofs are not always rigorous according to the standards of a later day, and his perspective slides back and forth with an ease that makes some of his investigations difficult to follow, but his conclusions remain basically sound even in those few situations where his analysis a strayed.

Laplace is remembered as one of the greatest scientists of all time. Sometimes referred to as the *French Newton* or *Newton of France*, he has been described as possessing a phenomenal natural mathematical faculty superior to that of any of his contemporaries. He died in Paris on March 5, 1827. He was 77 years old.

Laplace's name is one of the 72 names of the greatest scientists and engineers of France, placed on the first floor of the Eiffel Tower.

CHAPTER 11

MODERN MATHEMATICS (19TH-21ST CENTURIES)

The 19th century saw an unprecedented increase in the breadth and complexity of mathematical concepts. Both France and Germany were caught up in the age of revolution, which swept Europe in the late 18th century, but the two countries treated mathematics quite differently. After the French Revolution, Napoleon emphasized the practical usefulness of mathematics and his reforms and military ambitions gave French mathematics a big boost. Germany, on the other hand, under the influence of the great educationalist Wilhelm von Humboldt, took a rather different approach, supporting pure mathematics for its own sake, detached from the demands of the state and military (Ref 95).

The 20th century continued the trend of the 19th toward increasing generalization and abstraction in mathematics, in which the notion of axioms as "self-evident truths" was largely discarded in favor of an emphasis on such logical concepts as consistency and completeness. It also saw mathematics become a major profession, involving thousands of new Ph.D.'s each year and jobs in both teaching and industry, and the development of hundreds of specialized areas and fields of study, such as group theory, knot theory, sheaf theory, topology, graph theory, functional analysis, singularity theory, catastrophe theory, chaos theory, model theory, category theory, game theory, complexity theory and many more (Ref 96).

In 2000, the Clay Mathematics Institute[1] announced the seven Millennium Prize Problems, and in 2003 the Poincaré[2] conjecture was solved by Grigori Perelman[3] (who declined to accept an award, as he was critical of the mathematics establishment). Most mathematical journals now have online versions as well as print versions, and many online-only journals are launched. There is an increasing <u>drive towards open access publishing (Ref 2)</u>.

---

[1] The Clay Mathematics Institute (CMI) is a private, non-profit foundation, based in Peterborough, New Hampshire, United States.

[2] Henri Poincaré (1854–1912) was a French mathematician, theoretical physicist, and philosopher of science. He was responsible for formulating the Poincaré conjecture, which was one of the most famous unsolved problems in mathematics (About Poincaré and his work in mathematics are provided in Section 11.9).

[3] Grigori Perelman (born on June 13, 1966) is a Russian mathematician who made contributions to Riemannian geometry and geometric topology. (About Perelman and his work in mathematics are provided in Section 11.14).

11.1 Carl Friedrich Gauss (1777–1855)

Portrait of Carl Friedrich Gauss by Christian Albrecht Jensen (1840)

Carl Friedrich Gauss (Ref 97) was a German mathematician and physicist who contributed significantly to algebra, analysis, differential geometry, number theory, matrix theory, statistics, mechanics, electrostatics, magnetic fields, geophysics, astronomy, optics, and geodesy.

Carl Friedrich Gauss was born on April 30, 1777, in Brunswick, Germany. At the age of two, Carl showed himself to be a child prodigy. At 3 years old, he was able to read and write, even corrected his father's arithmetic. When Carl was in the third grade, the mathematics teacher J. G. Büttner gave his students the task: to calculate the sum of numbers from 1 to 100. He believed that solution to this problem would keep students busy until the end of the lesson. The teacher was amazed when Carl solved this problem within seconds and brought the correct result. Carl noticed that the pairwise sums from the opposite ends are the same: $1 + 100 = 101$, $2 + 99 = 101$, etc., and instantly got the result: $50 \times 101 = 5050$.

Carl Gauss developed a formula for finding the sum of any arithmetic progression. In modern mathematical notation his formula looks like this:

$$S_n = \frac{n}{2} \times (b + c)$$

where *n* is the number of terms, *b* is the first term, and *c* is the last term. Solving this problem by the Gauss method, we get:

$$S_n = 100/2 \times (1 + 100) = 50 \times 101 = 5050$$

Carl's intellectual abilities attracted the attention of Charles William Ferdinand, Duke of Brunswick, who sent him to the Collegium Carolinum (now Braunschweig University of Technology). Carl attended it from 1792 to 1795, and then he studied at the University of Göttingen from 1795 to 1798.

Gauss advanced number theory by discovering that every positive integer is representable as a sum of the triangular numbers. Such number counts objects arranged in an equilateral triangle, as in the diagram below.

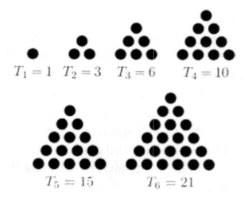

The *n*th triangular number is the number of dots composing a triangle with *n* dots on a side, and is equal to the sum of the *n* natural numbers from 1 to *n*. The *n*th natural number is calculated by the formula:

$$T_n = \frac{n(n+1)}{2}$$

The sequence of triangular numbers, starting at 0th triangular number, is:

1, 3, 6, 10, 15, 21, 28, 36, 45, <u>55</u>, 66, 78, 91, 105, <u>120</u>, 136, 153, 171, 190, 210...

Gauss was so excited by the results that, on July 10, 1796, he jotted down in his diary the note: EUREKA!

Let's check if the above formula is accurate:

$T_{10} = 10 \times (10 + 1)/2 = 110/2 = \underline{55}$
$T_{15} = 15 \times (15 + 1)/2 = 240/2 = \underline{120}$

As we see, the formula is accurate.

In 1796, Gauss showed that a regular polygon can be constructed by compass and ruler. This was a major discovery in an important field of mathematics. Construction problems had occupied mathematicians since the days of the Ancient Greeks, and the discovery ultimately led Gauss to choose mathematics instead of philology as a career. He was so pleased by this result that he requested that a regular heptadecagon (in geometry, a heptadecagon is a seventeen-sided polygon) be inscribed on his tombstone. The stonemason declined, stating that the difficult construction would essentially look like a circle. Regular heptadecagon is shown below.

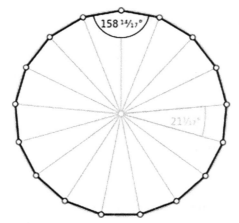

Heptadecagon (seventeen-sided polygon)

By the way, if you are interested in calculating internal (α) and vertical (β) angles of regular polygons, use the following formulas.

For internal angles: $\alpha = (n-2) \times 180/n$ (1)

Let $n = 17$, $\alpha = (17-2) \times 180/17 = 15 \times 180/17 = 158.8235°$

For vertical angles: $\beta = 360/n$ (2)

Let $n = 17$, $\beta = 360/17 = 21.1765°$

Keep in mind that the sum of these angles is: $\alpha + \beta = 158.8235° + 21.1765° = 180°$

There is another method to determine the values of internal angle of regular polygons. First, calculate the vertical angle using formula (2).

For example, the regular polygon is dodecagon (twelve-sided polygon). Its vertical angle is calculated by formula (2):

$\beta = 360/n = 360/12 = 30°$

The internal angle of this polygon is: 180° − 30° = 150°

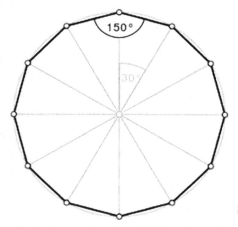

Dodecagon (twelve-sided polygon)

In 1798, Gauss accomplished *Arithmetical Investigations*, his magnum opus (Latin, *great work*). It was a textbook of number theory. This work was fundamental in built up number theory as a discipline and has shaped the field to the present day.

Gauss advanced modular arithmetic that greatly simplified manipulations in number theory.

A few words about modular arithmetic (Ref 98). When we divide two integers, we have an equation:

$A/B = Q$ remainder R, where:

A is the dividend
B is the divisor
Q is the quotient
R is the remainder.

Let's we are only interested in what the remainder is when we divide A by B. For this case there is an operator called modulo (abbreviated mod). Using the same A, B, Q, and R we have: $A \mod B = R$, or we say A modulo B is equal to R, where B is referred to as the modulus.

For example, $10/7 = 1$ remainder 3, $10 \mod 7 = 3$.

Modular arithmetic is referenced in number theory, group theory, ring theory, knot theory, abstract algebra, computer algebra, cryptography, computer science, chemistry and the visual and musical arts (Ref 99).

Although Gauss made contributions in almost all fields of mathematics (Ref 100), number theory was always his favorite area, and he asserted that "mathematics is the queen of the sciences, and the theory of numbers is the queen of mathematics." An example of how Gauss revolutionized number theory can be seen in his work with complex numbers—combinations of real and imaginary numbers. Gauss was not the first to interpret complex numbers graphically, but he was certainly responsible for popularizing the practice and also formally introduced the standard notation $a + bi$ for complex numbers. As a result, the theory of complex numbers received a notable expansion, and its full potential began to be unleashed.

In the area of probability and statistics, Gauss introduced what is now known as Gaussian distribution, the Gaussian function and the Gaussian error curve. He showed how probability could be represented by a bell-shaped or "normal" curve, which peaks around the mean or expected value and quickly falls off towards plus/minus infinity, which is basic to descriptions of statistically distributed data.

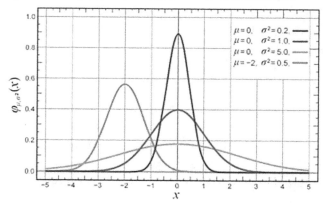

The normal or Gaussian curve of a probability distribution is used widely in statistics.
In the graph below, the red line is the normal curve, μ is the expected value and σ^2 is the variance.

The Kingdom of Hanover geodetic survey work in 1818, fueled Gauss' interest in differential geometry (a field of mathematics dealing with curves and surfaces) and what has come to be known as Gaussian curvature (an intrinsic measure of curvature, dependent only on how distances are measured on the surface, not on the way it is embedded in space).

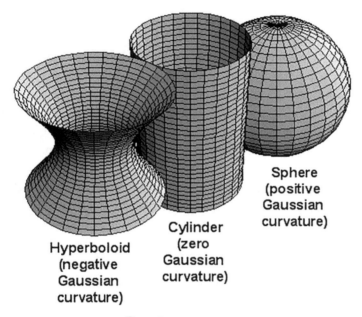

Gaussian curvatures

To aid the survey, Gauss invented the heliotrope, an instrument that uses a mirror to reflect sunlight over great distances, to measure positions.

Unfortunately for mathematics, Gauss reworked and improved his papers incessantly, therefore published only a fraction of them, in keeping with his motto "few but ripe." Many of his results were subsequently repeated by others since his terse diary remained unpublished for years after his death. This diary was only 19 pages long but later confirmed his priority on many results he had not published.

Gauss' achievements were not limited to mathematics. He was also a renowned physicist (Ref 85, pp. 11, 34, 157, 160–163, 284).

In 1832, Gauss promoted the application of the metric system. He was the first to make measurements of the Earth's magnetic force in terms of a decimal system based on the three mechanical units: *millimeter*, *gram*, and *second* for the quantities length, mass, and time.

In later years, Carl Friedrich Gauss and German physicist Wilhelm Eduard Weber (1804–1891) extended these measurements to include electrical concepts. At the time, they published a joint paper, which introduced absolute units of measurement of magnetism. It was the first publication on this subject.

In 1833, Gauss and Weber founded the Göttingen Magnetic Club and constructed the first battery operated electromagnetic telegraph line of 3-kilometer-long (about 1.9-mile-long), connecting the Physical Laboratory and Observatory at Göttingen. This telegraph allowed simultaneous magnetic observation at these sites. The Magnetic Club supported measurements of the Earth's magnetic field in many regions of the world.

In 1874, the British Association for the Advancement of Science introduced the CGS (centimeter-gram-second) system of units. Among derived units of this system are the *maxwell* (symbol Mx), a unit of magnetic flux across a surface of 1 cm^2 perpendicular to the magnetic field (named after James Clerk Maxwell, a British physicist and mathematician), and the *gauss* (symbol G), a unit of magnetic flux density (named after Carl Friedrich Gauss). Equivalently, one *gauss* represents a magnetic flux density of one *maxwell* per square centimeter area, 1 Mx = 1 G × cm^2.

This system is still in use, particularly by physicist worldwide, because many electromagnetic formulas are simpler in CGS (centimeter-gram-second) units.

Gauss died in Göttingen, on February 23, 1855. He was 77 years old.

11.2 Augustin-Louis Cauchy (1789–1857)

Portrait of Augustin-Louis Cauchy by Jean Roller (1840)

Augustin-Louis Cauchy (Ref 101) was a French mathematician and physicist who made pioneering contributions to mathematical analysis. He was one of the first to state and prove theorems of calculus rigorously, rejecting the heuristic principle of the generality of algebra of earlier authors. He almost singlehandedly founded complex analysis and the study of permutation groups in abstract algebra.

The Cauchy family survived the French Revolution (July 14, 1789) and the following Reign of Terror (1793–4) by escaping to Arcueil (a town located 5.3 km from the center of Paris) in July 1789. Augustin-Louis was born on August 21, 1789. In 1794, the Cauchy family returned to Paris.

In 1799, Napoleon Bonaparte came to power and Louis-François Cauchy (father of Augustin-Louis) was promoted to Secretary-General of the Senate, working directly under Pierre-Simon Laplace, famous mathematician. The other famous mathematician Joseph-Louis Lagrange was also a friend of the Cauchy family.

In the fall of 1802, on Lagrange's advice, Augustin-Louis was enrolled in the École Centrale du Panthéon (the School of the

Pantheon), the best secondary school of Paris at the time, where most of the curriculums consisted of classical languages. The young and ambitious Cauchy, being a brilliant student, won many prizes in Latin and Humanities. In spite of these successes, Augustin-Louis chose an engineering career, and prepared himself for entrance examination to the École Polytechnique, a French public institution of higher education and research. In 1805, Augustin-Louis passed the entrance examination, took the second place out of 293 applicants and was admitted. One of the main purposes of this school was to give future civil and military engineers a high-level scientific and mathematical education.

In 1807, Augustin-Louis Cauchy finished the École Polytechnique and went to the École des Ponts et Chaussées (School for Bridges and Roads). In 1810, Cauchy graduated from the School in specialty civil engineering with the highest honors. He accepted a job as a junior engineer in Cherbourg (city in the northwestern France) and stayed there for three years. Cauchy had an extremely busy managerial job, but he found time to prepare three mathematical manuscripts, which he submitted to the Institut de France (the Institute of France). Cauchy' first two manuscripts on *Polyhedrons* were accepted, but the third one on *Directrixes of conic sections* was rejected.

In September 1812, Cauchy returned to Paris, because he was losing his interest in engineering job, being more attracted to the abstract beauty of mathematics. The next three years Augustin-Louis was on unpaid sick leave and spent his time working on the *symmetric functions*, the *symmetric group*, and the *theory of higher-order algebraic equations*.

In 1815, Napoleon was defeated at Waterloo, and the newly installed Bourbon King Louis XVIII took the restoration in hand. The Académie des Sciences (Academy of Sciences) was re-established in March 1816, and the king appointed Cauchy to take one of two vacated places. The reaction by Cauchy's peers was harsh. They considered his acceptance of membership of the Academy an outrage, and Cauchy thereby created many enemies in scientific circles.

However, Cauchy was a rising mathematical star, who merited a professorship. One of his great successes at that time was the proof of Fermat's polygonal number theorem. The Fermat polygonal number theorem states that every positive integer is a sum of n-gonal numbers. That is, every positive integer can be written as the sum of three or fewer triangular numbers, and as the sum of four or fewer square numbers, and as the sum of five or fewer pentagonal numbers, and so on. For example, three such representations of the number 17 are:

$17 = 10 + 6 + 1$ (*triangular numbers*)
$17 = 4^2 + 1^2$ (*square numbers*)
$17 = 12 + 5$ (*pentagonal numbers*).

Cauchy finally quit his engineering job and received a one-year contract for teaching mathematics to second-year students of the École Polytechnique. In 1816, this non-religious school was reorganized, and several liberal professors were fired. The reactionary Cauchy, at the age of 27 years old, was promoted to full professor. The conservative political climate lasted until 1830. During that period, Cauchy was highly productive and published several important mathematical treatises.

The French Revolution of 1830 (also known as the July Revolution) occurred. Riots, in which students of the École Polytechnique took an active part, raged close to Cauchy's home in Paris. These events marked a turning point in Cauchy's life and broke in his work on mathematics.

In 1831, Cauchy went to the city of Turin in Italy, and accepted an offer from the King of Sardinia for a chair of theoretical physics, which was created especially for him. He taught in Turin during 1832–33. In 1831, he was elected a foreign member of the Royal Swedish Academy of Sciences, and the following year—a foreign honorary member of the American Academy of Arts and Sciences.

In late 1838, Cauchy returned to Paris and reinstated his post at the Academy of Sciences. He could not regain his teaching position, because he refused to swear an oath allegiance. However, he desperately wanted to regain a formal position to be engaged in science.

Cauchy wrote over 800 papers. Complete collection of his works contains 27 volumes. He was very productive in number of papers—second only to Leonhard Euler. Cauchy's work relates to various fields of mathematics and mathematical physics.

Cauchy gave a strict definition to the basic concepts of mathematical analysis: limit, continuity, derivative, differential, integral, convergence of series, etc. His definition of continuity was based on the concept of infinitesimal that it is a variable quantity, which tends to zero. He introduced the concept of the radius of convergence of a series. The Cauchy analysis courses, based on the systematic use of the concept of limit, served as a model for most courses of the later period.

Cauchy worked a lot in the field of complex analysis, in particular, he created the theory of integral residues. In mathematical physics he deeply studied the boundary-value problem with initial conditions, which has since been called the "Cauchy problem." He also did research on polyhedra (a solid figure with many plane faces, typically more than 6), on number theory, algebra, and other areas of mathematics.

Cauchy died of bronchial condition on May 23, 1857. He was 67 years old. His name is one of the 72 names of the greatest scientists and engineers of France, placed on the first floor of the Eiffel Tower.

11.3 Nikolai Ivanovich Lobachevsky (1792–1856)

Portrait of Nikolai Lobachevsky by Lev Kryukov (c. 1843)

Nikolai Ivanovich Lobachevsky (Ref 102) was a Russian mathematician, best known for his work on hyperbolic geometry (in mathematics, hyperbolic geometry is a non-Euclidean geometry, also called Lobachevskian geometry).

Nikolai was born near the city of Nizhniy Novgorod in the Russian Empire on December 1, 1792, to parents Ivan Maksimovich Lobachevsky and Praskovia Aleksandrovna Lobachevskaya. Nikolai was one of three children. His father died in 1799, and his mother moved to the city of Kazan (currently, Kazan is the capital and largest city of the Republic of Tatarstan, Russia).

Nikolai Lobachevsky attended Kazan Gymnasium from 1802, graduating in 1807 and then received a scholarship to Kazan University, which was founded just three years earlier in 1804. At Kazan University, Lobachevsky was influenced by Professor Johann Bartels, a former teacher and friend of German mathematician Carl Friedrich Gauss. In 1811, Lobachevsky received a master's degree in physics and mathematics. In 1814, he became a lecturer at Kazan University, in 1816, he was promoted to associate professor, and in 1822, being 30 years old, he became a full professor, teaching mathematics, physics, and astronomy. He served in many administrative positions and became the rector of Kazan University in 1827.

In 1832, Lobachevsky married Varvara Alekseevna Moiseeva, who was 20 years younger than her husband. They had 18 children, but only 7 survived into adulthood.

Lobachevsky's main achievement is the development of a non-Euclidean geometry (also referred to as Lobachevskian geometry). Before him, mathematicians were trying to deduce Euclid's fifth postulate: "That, if a straight line falling on two straight lines make the interior angles on the same side less than two right angles, the two straight lines, if produced indefinitely, meet on that side on which are the angles less than the two right angles."

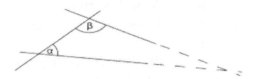

Lobachevsky would instead develop a geometry in which the fifth postulate was not true. On February 23, 1826, he reported his research to the session of the department of physics and mathematics, and it was printed in the Bulletin of Kazan University. In 1829–1830, Lobachevsky wrote a paper called "A concise outline of the foundations of geometry" that was published by the Kazan Messenger but was rejected when it was submitted to the St. Petersburg Academy of Sciences for publication.

Lobachevsky developed the non-Euclidean geometry that is referred to as hyperbolic geometry (also called Lobachevskian geometry). The parallel postulate of Euclidean geometry is replaced with: "for any given line R and point P not on R, in the plane containing both line R and point P there are at least two distinct lines x and y through P that do not intersect R", as shown bellow.

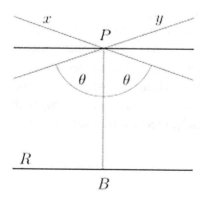

A visual representation of the geometry of Lobachevsky: through the point M there pass two lines parallel to the line D.

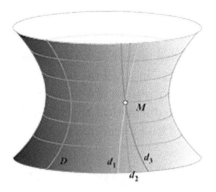

Non-Euclidean geometry stimulated the development of differential geometry, which has many branches. One of them is Riemannian geometry. It is named after Friedrich Bernhard Riemann, a German mathematician. Detailed information about him is given in Section 11.7.

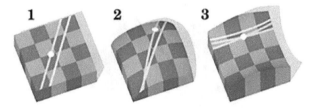

1. Euclidean geometry 2. Riemannian geometry 3. Lobachevskian geometry

On April 18, 1845, Lobachevsky became a trustee of the Kazan academic district. On November 20, 1845, Lobachevsky was elected unanimously for the sixth time as a rector of Kazan University for a new four-year period. In 1846, after 30 years of service, the ministry, according to the statute, was to decide whether to leave Lobachevsky as professor or to choose a new teacher. The university council informed the minister that there is no reason to remove Lobachevsky from teaching.

Lobachevsky was dismissed from the university in 1846, due to his deteriorating health: by the early 1850s, he was nearly blind and unable to walk. He died in poverty in 1856 at 63 years old. He was unrecognized in his country, not having lived until the triumph of his ideas for only 10-12 years.

Lobachevsky obtained a number of valuable results in other branches of mathematics: in algebra, he developed a method for approximate solution of equations, obtained a number of theorems on trigonometric series in mathematical analysis, refined the notion of a continuous function, and gave a criterion for the convergence of series.

11.4 Niels Henrik Abel (1802–1829)

Niels Henrik Abel (Ref 103) was a Norwegian mathematician who made pioneering contribution in a variety of fields. His most famous result is the proof demonstrating the impossibility of solving the general quintic equation in radicals. In mathematics, a radical expression is defined as any expression containing a radical ($\sqrt{}$) symbol to determine the square root ($x = 2$) of a number (y).

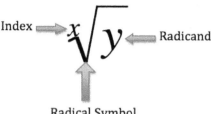

Radical Symbol

In 1815, Niels Abel entered the Cathedral School at the age of 13. His elder brother Hans joined him there a year later. They shared rooms and had classes together. Hans got better grades than Niels.

In 1818, a new mathematics teacher Bernt Michael Holmboe gave the students mathematical tasks to do at home. He quickly discovered Abel's talent in mathematics and proclaimed him as a "splendid genius" in his report card. Holmboe encouraged him to study the subject to an advanced level, and even gave Niels private lessons after school. Holmboe supported Niels with a scholarship to remain at the school and raised the money to enable him to study at the Royal Frederik University.

When Abel entered the Royal Frederik University in 1821, he was already the most knowledgeable mathematician in Norway. He had studied all the latest mathematical literature in the university library. During that time, he started working on the quintic equation in radicals.

In algebra, the Abel impossibility theorem states that there is no algebraic solution in radicals of degree five or higher with arbitrary coefficients. The theorem is named after Niels Henrik Abel, who provided a proof in 1824.

The general quintic equation has the form: $ax^5 + bx^4 + cx^3 + dx^2 + ex + f = 0$.

The degree four (quartic case) is the highest degree such that every polynomial equation can be solved by radicals.

The general quadratic equation has the form: $ax^2 + bx + c = 0$. Solution of that equation is described in Section 6.3 of this book.

The general cubic equation has the form: $ax^3 + bx^2 + cx + d = 0$. Formulas for solution of the cubic equation are:

$$C = \sqrt[3]{\frac{\Delta_1 \pm \sqrt{\Delta_1^2 - 4\Delta_0^3}}{2}}.$$

Solution of the cubic equation involves first calculating:

$\Delta_1 = 2b^3 - 9abc + 27a^2d$
$\Delta_0 = b^2 - 3ac$

There are three possible cube roots implied by the expression, of which at least two are non-real complex numbers; any of these may be chosen when defining C.

The general formula for one of the roots, in terms of the coefficients, is:

$$x = -\frac{1}{3a}\left(b + C + \frac{\Delta_0}{C}\right).$$

The general quartic (the degree four) equation has the form:

$ax^4 + bx^3 + cx^2 + dx + e = 0$

Solution of the general quartic equation requires many more formulas than that for the general cubic equation. Those, who are interested in it, can visit: *en.wikipedia.org/wiki/Quartic function*.

Abel wrote a fundamental work on the theory of elliptic integrals, containing the foundations of the theory of elliptic functions. Historically, elliptic functions were first discovered by Abel as inverse functions of elliptic integrals, and their theory was improved by Carl Gustav Jacob Jacobi (1804–1851), a German mathematician. Jacobi was the first Jewish mathematician to be appointed professor at a German university.

In the winter of 1828, Abel contracted tuberculosis. At Christmas 1828, he traveled by sled to Froland (a municipality in Aust–Agder county, Norway) to visit his fiancée. He became seriously ill on the journey, although a temporary improvement allowed the couple to enjoy the holiday together. Niels Henrik Abel died on April 6, 1829. He was 26 years old.

Under Abel's guidance, the prevailing obscurities of analysis (that is the process of breaking a complex topic or substance into smaller parts in order to gain a better understanding of it) began to be cleared, new fields were entered upon and the study of functions so advanced as to provide mathematicians with numerous ramifications along with progress could be made.

There are 20 titles in mathematical writing named after Niels Henrik Abel, such as *Abel's binomial theorem*, *Abelian category*, *Abelian equation*, *Abelian group*, *Abel's inequality* and so on, including Abel Prize.

The Abel's work was done in six or seven years of his working life. Regarding him, the French mathematician Adrien-Marie Legendre (1752–1833), said, "What a head the young Norwegian has!"

In August 2001, the Norwegian government announced that the prize would be awarded beginning in 2002, the two-hundredth anniversary of Abel's birth. Abel Prize comes with a monetary award of 6 million Norwegian Kroner ($700,000). The first Abel Prize was awarded in 2003. From 2003 to 2020, 22 outstanding mathematicians were awarded the Abel Prize. The first winner is Jean-Pierre Serre (citizenship – France) and in 2020, the winners are Hillel Furstenberg (citizenships – Israel and United States) and Gregory Margulis (citizenships – Russia and United States). These 22 winners of the Abel Prize were represented by 9 countries, surprisingly Norway was not amongst them.

11.5 Évariste Galois (1811–1832)

Évariste Galois (Ref 104) was a French mathematician. While still in his teens, he was able to determine a necessary and sufficient condition for a polynomial to be solvable by radicals. His work laid the foundations for Galois theory and group theory, which are the two major branches of abstract algebra, and the subfield of Galois connections.

At the age of 17, Évariste Galois (Ref 105) began making fundamental discoveries in the theory of polynomial equations (equations constructed from variables and constants, using only the operations of addition, subtraction, multiplication and non-negative whole-number exponents, such as $x^2 - 4x + 7 = 0$).

Galois proved that there can be no general formula for solving quintic equations (polynomials including a term of x^5), just as the young Norwegian mathematician Niels Henrik Abel did a few years earlier, although by a different method.

Galois also proved that there is no general algebraic method for solving polynomial equations of any degree greater than four. The Galois proof was more general and more powerful than that by Abel. Galois achieved this general proof by looking at whether or not the "permutation group" of its roots (now known as its Galois group) had a certain structure. He was the first to use the term "group" in its modern mathematical sense of a group of permutation, and his fertile approach, now known as Galois Theory, was adapted by later mathematicians to many other fields of mathematics besides the theory of equations.

Note: in mathematics, the permutation relates to the act of arranging all the members of a set into some sequence or order or if the set is already ordered, rearranging its elements its elements, a process called permuting.

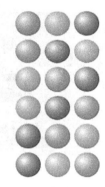

Each of the six rows is a different permutation of three distinct balls.

Galois was a hot-headed political firebrand (he was arrested several times for political acts), and his political affiliations and activities as a staunch republican during the rule of Louis Philippe I[4]).

Galois was continually distracted from his mathematical work. He died from wounds suffered in a duel in 1832, at age 20, under rather shady circumstances. He had spent the whole of the previous night outlining his mathematical ideas in a detailed letter to his friend Auguste Chevalier, as though convinced of his impending death.

There are 20 titles in mathematical writing named after Évariste Galois, among which are *Absolute Galois group, Differential Galois theory, Galois closure, Galois connection, Galois covering,* and so on.

---

[4] Louis Philippe I (1773–1850) was King of the French from 1830 to 1848 and the last French king.

Within the 60 or so pages of Galois' collected works are many important ideas that have had far-reaching consequences for nearly all branches of mathematics. His work has been compared to that of Niels Henrik Abel, another mathematician who died at a very young age, and much of their work had significant overlap.

11.6 János Bolyai (1802–1860)

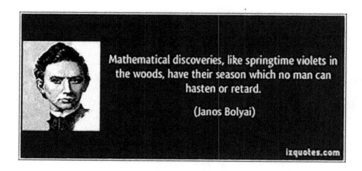

János Bolyai (Ref 106) was a Hungarian mathematician, one of the founders of a geometry that differs from Euclidean geometry in its definition of parallel lines. The discovery of a consistent alternative geometry that might correspond to the structure of the universe helped mathematicians to study abstract concepts irrespective of any possible connection with the physical world.

János was born on December 15, 1802, in the Transylvanian town of Kolozsvár (now Cluj-Napoca in Romania). His father, Farkas Bolyai was well-known mathematician. By the age of 13, János had mastered calculus and other forms of analytical mechanics, receiving instruction from his father. János studied at the Theresian Military Academy from 1818 to 1822. He graduated from the Academy having passed the seven-year course for 4 years.

During his study, János became so obsessed with Euclid's parallel postulate (see Section 4.5 of this book) that his father, who pursued the same subject for many years, wrote to him in 1820: "You must not attempt this approach to parallels. I know this way to the very end. I have traversed this bottomless night, which extinguished all light and joy in my life. I entreat you, leave the science of parallels alone...Learn from my example."

János Bolyai persisted in his quest and eventually came to conclusion that Euclid's fifth postulate is independent of the other axioms of his geometry. Between 1820 and 1823, he had prepared a treatise on a complete system of non-Euclidean geometry, and his work was published in 1832 as an appendix to a mathematics textbook by his father.

In 1848, Bolyai discovered that Russian mathematician Nikolai Ivanovich Lobachevsky had published a similar work in 1829. They did not know each other or each other's works. Their works contained hyperbolic geometry. In mathematics, hyperbolic geometry (which also called Bolyai–Lobachevskian geometry or Lobachevskian geometry) is a non-Euclidean geometry.

In addition to his work in geometry, Bolyai developed a rigorous geometric <u>concept of complex numbers[5]) as ordered pairs of real numbers[6])</u>.

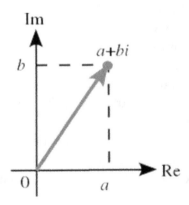

"Re" is the real axis, "Im" is the imaginary axis, and i satisfies $i^2 = -1$

---

[5] A complex number is any number that can be written as $a + bi$, where i is the imaginary unit, a, b are real numbers, and i is a solution of the equation $x^2 = -1$. Because no real number satisfies this equation, i is called an imaginary number.

[6] A real number is a value of a continuous quantity that can represent a distance along a line.

A symbol of the set of real numbers

In recent years, Bolyai had a serious mental disorder. He died of pneumonia on January 27, 1860, at the age of 57. He left more than 20,000 pages of mathematical manuscripts. János Bolyai was an accomplished polyglot speaking nine languages, including Chinese and Tibetan.

Several universities, many primary and secondary schools in Hungary and Rumania are named after János Bolyai. The International János Bolyai Prize of Mathematics is an international prize founded by the Hungarian Academy of Sciences. The prize is awarded every five years to mathematicians for monographs with important new results in the preceding 10 years.

The award was founded in 1902 and was awarded twice before the First World War. Due to the outbreak of the war, the third award in 1915 did not take place. The award was given to the scientist who had the greatest influence on the development of mathematics over the past 25 years and amounts to 10 thousand crowns.

The award was revived in 1994. Awards have been held every five years since 2000. The prize is awarded for the best monographs in mathematics over the past 15 years and includes $25,000 and a bronze medal.

From 1902 to 2015, six mathematicians were awarded the János Bolyai Prize:

      1905, Henri Poincaré, France
      1910, David Hilbert, Germany
      2000, Saharon Shelah, Israel
      2005, Mikhail Leonidovich Gromov, Russia, France

2010, Yuri Ivanovich Manin, Russia, Germany
2015, Barry Simon, USA

In 2020, this prize was awarded to Terence Chi-Shen Tao (born 17 July 1975). He is an Australian-American mathematician, professor of mathematics at the University of California, Los Angeles, holds the James and Carol Collins chair.

11.7 Bernhard Riemann (1826–1866)

Georg Friedrich Bernhard Riemann (Ref 107) was a German mathematician. His father was a poor Lutheran pastor. His mother died before her children had reached adulthood. Bernhard was the second of six children, shy and suffering from numerous nervous breakdowns. He exhibited exceptional mathematical skills, such as calculation abilities from an early age, but suffered from timidity and fear of speaking in public.

During 1840, Riemann attended middle school. In 1842, he attended high school and studied the Bible intensively, but was often distracted by mathematics. His teachers were amazed by his adept ability to perform complicated mathematical operations, in which he outstripped his instructor's knowledge. At the age of 19, he started studying philosophy and Christian theology in order to become a pastor and help with his family's finances.

In 1846, Riemann was sent to the University of Göttingen, where he planned to study towards a degree in Theology. However, he began studying mathematics under Carl Friedrich Gauss. Gauss recommended that Riemann give up his theological work and enter the mathematical field. After getting his father's approval, Riemann transferred to the University of Berlin in 1847. During his time of study, the best German mathematicians were teaching differential equations, number theory, geometry, the theory of Fourier series, and mathematical analysis. Riemann stayed in Berlin for two years and returned to Göttingen in 1849.

Riemann held his first lectures in 1854, which founded the field of Riemannian geometry. In 1857, there was an attempt to promote Riemann to extraordinary professor status at the University of Göttingen. However, he was not approved.

In 1859, following Dirichlet's[7] death, Riemann was promoted to head the mathematics department at the University of Göttingen.

In 1859, Riemann was elected a member of the Berlin Academy of Sciences. In the following year, Riemann was elected a member of the Royal Society of London and a member of the Paris Academy of Sciences.

Riemann's published works opened up researched areas combining analysis with geometry. These would become major parts of the theories of Riemannian geometry, algebraic geometry, and complex

---

[7] Peter Gustav Lejeune Dirichlet (1805–1859) was a German mathematician who made deep contribution to number theory and to the theory of Fourier series[8].

[8] In mathematics, a Fourier series is a way to represent a function as the sum of simple sine waves. This series, the Fourier transform, and Fourier's law are named in the honor of a French mathematician and physicist Joseph Fourier (1768–1830).

manifold theory. The theory of Riemann surfaces <u>was elaborated by Felix Klein[9]) and Adolf Hurwitz[10])</u>.

Riemann contributed a lot to analysis, number theory, and differential geometry. He is known for the first rigorous formulation of the integral named after him (the Riemann integral). He contributed to complex analysis by the introduction of Riemann surfaces. Riemann's famous 1859 paper on the prime-counting function, containing the original statement of the Riemann hypothesis, is regarded as one of the most influential papers in analytic number theory. Through his pioneering contributions to differential geometry, Riemann laid the foundations of the mathematics of general relativity.

Academician Pavel Sergeevich Aleksandrov (1896–1982), a mathematician of the Soviet Union, noted, "We tend to see in Riemann, perhaps, the greatest mathematician of the mid-19th century, the direct successor to Gauss."

In 1866, Riemann died of tuberculosis during his third journey to Italy. He was 39 years old.

---

[9] Felix Klein (1849–1925) was a German mathematician, known for his work with group theory, complex analysis, non-Euclidean geometry, and on the associations between geometry and group theory.

[10] Adolf Hurwitz (1859–1919) was a German mathematician who worked on algebra, analysis, geometry, and number theory. Since 1892, he taught mathematics at the Polytechnic School in Zurich. Among his students was Albert Einstein.

11.8 Sofia Vasilyevna Kovalevskaya (1850–1891)

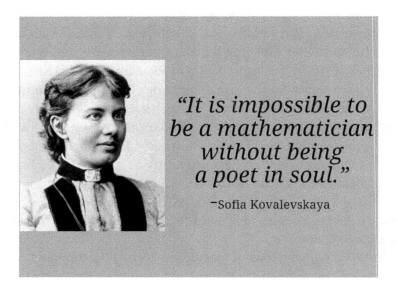

"It is impossible to be a mathematician without being a poet in soul."

−Sofia Kovalevskaya

Sofia Vasilyevna Kovalevskaya (Ref 108) was a Russian mathematician. She was born in Moscow on January 15, 1850, the second of three children. Her father Vasily Vasilyevich Korvin-Krukovsky served as Lieutenant General in the Imperial Russian Army as head of the Moscow Artillery.

Sofia's parents provided her with a good early education. Acquaintance of Sofia with mathematics occurred in early childhood. The walls of her room had been papered (by chance, due to lack of wallpaper) with notes by Ostrogradsky[11] on differential and integral calculus (left over from her father's student days).

From 8 years old, Sofia was tutored privately in elementary mathematics by Iosif Ignatievich Malevich. Many years later, he placed in "Russian Antiquity" magazine (December 1890) the article on his pupil.

When Sofia was fourteen years old (Ref 109), she taught herself trigonometry in order to understand the optics section of a physics

---

[11] Mikhail Vasilyevich Ostrogradsky (1801–1862) was a Russian mathematician.

book that she was reading. The author of the book and also her neighbor, Professor Nikolai Nikanorovich Tyrtov (1822–1888) extremely impressed with her capabilities and convinced her father to allow her to go off to school in St. Petersburg to continue her studies. After concluding her secondary schooling, Sofia was determined to continue her education at the university level.

Despite Sofia's talent for mathematics (Ref 108), she could not complete her education in Russia. At that time, women were not allowed to attend universities in Russia and most other countries. In order to study abroad, Sofia needed written permission from her father (or husband). In 1868, she contracted a "fictitious marriage" with Vladimir Kovalevsky (1842–1883), paleontologist and book publisher. He was the first who translated and published the works of Charles Darwin in Russia. They moved from Russia to Germany in 1869, in order to pursue advanced studies.

In April 1869, Sofia and Vladimir moved to Heidelberg. Through great efforts, Sofia obtained permission to audit classes at the University of Heidelberg[12] with the professors' approval.

Sofia attended courses in physics and mathematics under such teachers as Hermann von Helmholtz (1821–1894), Gustav Kirchhoff (1824 –1887), and Robert Bunsen[13] (1811–1899).

In October 1870, Sofia moved to Berlin, where she took private lessons by Karl Weierstrass (1815–1897), a German mathematician, because the university did not even allow her to audit classes. Karl Weierstrass was very impressed with her mathematical skills, and over the subsequent three years he taught her the same material that comprised his lectures at the university. He continued to teach Sofia for three more years.

In 1874, Sofia Kovalevskaya presented to the University of Göttingen three papers as her doctoral dissertation: "On partial

---

[12] Heidelberg University was founded in 1386. It is Germany's oldest university, and one of the world's oldest surviving universities.

[13] One of his doctoral students was Dmitry Mendeleyev (1834 –1907) from Russia. He was a chemist and inventor. Mendeleev formulated the Periodic Law and created a farsighted version of the periodic table of elements.

differential equations," "On elliptic integrals," and "On the dynamics of Saturn's rings."

With the support of Weierstrass, doctoral advisor of Sofia, she earned a doctorate degree in mathematics *summa cum laude*.

Sofia Kovalevskaya became the first woman to have been awarded a doctorate at a European university. Her paper "On partial differential equations" contains what is now commonly known as the Cauchy–Kovalevskaya theorem, associated with Cauchy initial value problems. A special case was proven by Augustin Cauchy (1842), and the full result by Sofia Kovalevskaya (1875).

In 1874, Sofia and her husband Vladimir returned to Russia. Vladimir failed to secure a professorship, because of his radical beliefs. A year later, for some unknown reason, Sofia and Vladimir decided to spend several years together as an actual married couple. Three years later, Sofia gave birth to a daughter Sofia (pet name "Fufa"). After two years devoted to raising her daughter, Kovalevskaya put Fufa under the care of relatives and friends, resumed her work in mathematics, and left Vladimir for what would be the last time.

In 1881, Sofia Kovalevskaya was elected to the Moscow Mathematical Society in the post of privat-docent.

After the suicide of Sofia's husband Vladimir (1883), who became entangled in his business affairs, she and her a five-year-old daughter went to Berlin without money and stopped at Karl Weierstrass. At the cost of enormous efforts, using all his authority and connections, Weierstrass managed to obtain her a place at the University of Stockholm. Kovalevskaya was admitted as a professor in the Department of Mathematics at the University of Stockholm. She had to give lectures the first year in German, and the second year in Swedish. Soon she mastered the Swedish language and was able to print in Swedish her mathematical work.

In 1884, Kovalevskaya was appointed to a five-year as Extraordinary Professor (assistant professor in modern terminology) and became an editor of *Acta Mathematica* (a scientific journal covering research in all fields of mathematics.

In 1888, Kovalevskaya won the *Prix Bordin* (Borden Prize) of the Paris Academy of Sciences for her discovery the third classical case of solvability of the problem of the rotation of a fixed point. Next year the second work on the same subject, she was awarded the prize of the Swedish Academy of Sciences.

In 1889, Sofia Kovalevskaya was appointed Ordinary Professor (full professor) at Stockholm University, the first woman in Europe to hold such a position. After much lobbing on her behalf (and a change in the Academy's rules) she was elected a Corresponding Member of the Russian Academy of Sciences but was never offered a professorship in Russia.

Sofia Kovalevskaya died of epidemic influenza complicated by pneumonia in 1891 at age forty-one. She is buried in Solna, Sweden, at the "Northern Cemetery" in Stockholm urban area.

She made significant contributions to mathematical analysis, partial differential equations and mechanics. She was a pioneer for women in mathematics around the world.

11.9 Henri Poincaré (1854–1912)

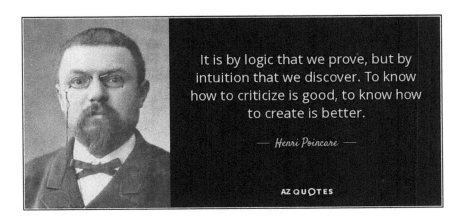

Henri Poincaré (Ref 110) was a French mathematician, theoretical physicist, engineer, and philosopher of science. He is often described as a polymath, and in mathematics as "The Last Universalist" since he excelled in all fields of that discipline as it existed during his lifetime.

Henri was born on April 29, 1854, into an intelligent and influential family. They lived in a neighborhood of Nancy. His father Leon Poincaré (1828–1892) was a professor of medicine at the University of Nancy (the city of Nancy is 283 kilometers west of Paris). Henri was raised in the Roman Catholic faith, but later left the religion. During his childhood, he was seriously ill for a time with diphtheria.

In 1862, Henri entered the Lycée in Nancy. He spent eleven years at this school. During this time he proved to be one of the top students in every topic he studied. His mathematics teacher described him as a "monster of mathematics." He won first prizes in the Concours Général. Across France, the Concours Général is the most prestigious academic competition held every year between the top students of 11th grade and 12th final grade in almost all subjects taught in general, technological, and professional high schools.

Henri's poorest subjects were music and physical education, where he was described as "average at best." His tendency towards absent-mindedness and poor eyesight may explain these difficulties. In 1871, Henri graduated from the Lycée with a bachelor's degree in letters and science.

In 1873, Henri Poincaré entered the École Polytechnique and graduated in 1875. There he studied mathematics and published his first paper the *New Demonstration of Indicator Properties of a Surface* in 1874. From November 1875 to June 1878, Poincaré studied at the School of Mines, while continuing the study of mathematics in addition to the mining engineering syllabus. In March 1879, he received the degree of ordinary mining engineer.

At the same time, Henri Poincaré was preparing for his doctorate in mathematics. His doctoral thesis was in the field of differential equations named "*On the properties of the functions defined by the partial difference equations.*" Poincaré received his doctorate in mathematics from the University of Paris in 1879. He devised a new way of studying the properties of these equations. Poincaré was the first person to study their general geometric properties. He realized that these equations could be used to model the behavior of multiple bodies in free motion within the solar system.

In December 1879, Poincaré began teaching as a junior lecturer in mathematics at the University of Caen in Normandy. At the same time he published his first major <u>article concerning the treatment of a class of automorphic functions</u>[14]).

In 1881–1882, Poincaré created a new branch of mathematics: qualitative theory of differential equations. He showed how it is possible to derive the most important information about the behavior of a family of solutions without having to solve the equation. He successfully used this approach to problems in celestial mechanics and mathematical physics.

In 1881, Poincaré accepted the invitation to take a teaching position at the Faculty of Sciences of the University of Paris. From 1883 to 1897, he taught mathematical analysis in École Polytechnique of Paris.

In 1885, the King of Sweden Oscar II organized a mathematical contest and offered participants a choice of four topics. The most difficult was the first: to calculate the movement of the gravitating bodies of the solar system. Poincaré showed that this problem (the so-called three-body problem) does not have a complete mathematical solution. Nevertheless, Poincaré soon proposed effective methods for its approximate solution. In 1889, he received the Swedish competition award. One of the two judges wrote about the work of Poincaré, "The prized memoir will be among the most significant mathematical discoveries of the century." The second judge said that after the work of Poincaré, "A new era will begin in the history of celestial mechanics." For this success, the French government awarded the Order of the Legion of Honor to Henri Poincaré.

In the fall of 1886, 32-year-old Poincaré headed the Department of Mathematical physics and Probability theory at the University of Paris. The symbol of recognition of Poincaré as the leading mathematician of France was his election as President of the French Mathematical Society (1886) and a member of the Paris Academy of Sciences (1887).

---

[14] In mathematics, an automorphic function is a function on a space that is invariant under the action of some group.

In 1887, Poincaré generalized several complex variables of the Cauchy[15] theorem <u>and initiated the theory of residues in a multidimensional complex space.</u>

In 1889, the fundamental "Course of Mathematical Physics" by Poincaré was published in 10 volumes, and in 1892–93, two volumes of the monograph "New methods of celestial mechanics" (the third volume was published in 1899).

Since 1893, Poincaré is a member of the prestigious Bureau of Longitude (in 1899 he was elected its president). In 1896, he transferred to the university department of celestial mechanics, which he held until the end of his life. In the same period, continuing to work on astronomy, he implements a long-though out plan for <u>creating high-quality geometry—topology[16]).</u>

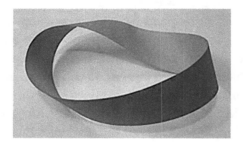

In 1906, Poincaré was elected President of the Paris Academy of Sciences.

In 1908, Poincaré became seriously ill and was unable to read his report "The Future of Mathematics" at the Fourth Mathematical Congress.

---

[15] About Cauchy and his contributions to mathematics are provided in Section 11.2

[16] Unlike geometry, the topology does not consider the metric properties of objects (for example, distance between a pair of points). In terms of topology, a cup and a bagel are indistinguishable. Möbius strips, which have only one surface and one edge, are a kind of object studied in topology. From 1894, Poincaré began publishing articles devoted to building a new, extremely promising science.

The first operation ended successfully, but after 4 years, his condition worsened again. He died in Paris after surgery from an embolus on July 17, 1912, at the age of 58 years. He was buried in the family crypt at the cemetery of Montparnasse.

The mathematical activity of Poincaré was of an interdisciplinary nature, so that in his thirty-plus years of intense work he left fundamental works in almost all areas of mathematics. The works of Poincaré, published by the Paris Academy of Sciences in 1916–1956, comprise 11 volumes. His works are on the topology, automorphic functions, theory of differential equations, multidimensional complex analysis, integral equations, non-Euclidean geometry, probability theory, number theory, celestial mechanics, physics, philosophy of mathematics, and philosophy of science.

In all the various fields of his work, Poincaré obtained important and profound results. In his scientific heritage there are a lot of major works on "pure mathematics" (general algebra, algebraic geometry, number theory, etc.). The results of his works have direct applied application and still significantly dominate. This is especially noticeable in his works of the last 15-20 years. Nevertheless, the discoveries of Poincaré, as a rule, were of a general nature and were later successfully applied in other fields of science.

The creative method of Poincaré relied on the creation of an intuitive model of the problem posed: he always first completely solved the problems in his head and then wrote down the solution. Poincaré had a phenomenal memory and could quote word by word from the books read and conversations held.

The Poincaré conjecture "Every simply connected, closed 3-manifold is homeo-morphic to the 3-sphere," during his lifetime was not proven. One hundred years have passed, and Grigori Perelman proved this conjecture in 2003.

About Grigori Perelman and his contributions to mathematics are provided in Section 11.14.

11.10 Georg Cantor (1845–1918)

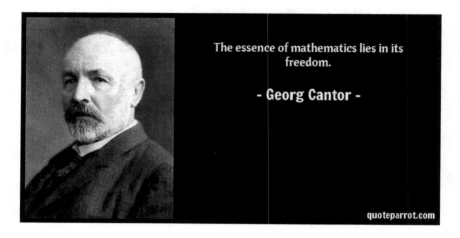

Georg Cantor (Ref 111) was a German mathematician. He created set theory, which has become a fundamental theory in mathematics.

Georg was born on March 3, 1845, in the Western merchant colony in St. Petersburg, Russia. His father, Georg-Waldemar Cantor came from Portuguese Jews who had settled in Amsterdam and was a Danish citizen. Georg Cantor grew up in St. Petersburg until he was eleven years old.

Georg was the eldest of six children. He masterfully played the violin, inheriting from his parents considerable artistic and musical talents. The father of the family was a member of the St. Petersburg Stock Exchange. When he fell ill, the family, counting on a milder climate, moved to Germany in 1856: first to Wiesbaden, and then to Frankfurt.

In 1860, Georg graduated with honors from the Darmstadt School. Teachers noted his exceptional ability in mathematics, in particular, in trigonometry. In 1862, he entered the Federal Polytechnic Institute in Zurich. A year later, his father died. Having received a substantial legacy, Georg transferred to the Humboldt University of Berlin. He spent the summer of 1866, at the University of Göttingen—the largest center of mathematical thought of those times. In 1867, the University of Berlin awarded Georg Cantor a Ph.D. degree for his work on number theory.

After teaching briefly at the Berlin School for Girls, Cantor took a place at the Martin Luther University of Halle-Wittenberg, where he spent his entire career.

Cantor was awarded the requisite habilitation for his thesis and work on number theory, which he presented in 1869 upon his appointment at University of Halle-Wittenberg.

In 1872, Cantor was promoted to extraordinary professor and became full professor in 1879. To attain such rank at the age of 34 was a notable accomplishment, but he desired a chair at a more prestigious university, in particular at Berlin, at that time the leading German university. However, his work encountered too much opposition for that to be possible.

In 1877, Cantor obtained an astounding result that he reported to his friend Richard Dedekind[17] in a letter: "The sets of points of a segment and points of a square have the same power (continuum), regardless of the length of the segment and the width of the square."

Cantor formulated and unsuccessfully tried to prove the "continuum hypothesis". His first article outlining these key results appeared in 1878 and was called "Toward the Doctrine of Diversity" (he replaced the term "diversity" later by "many"). The publication of the article was repeatedly postponed at the request of the indignant Leopold Kronecker[18], who headed the mathematics department of the University of Berlin.

---

[17] Richard Dedekind (1831–1916) was a German mathematician who made excessive contributions to abstract algebra, algebraic number theory, and the definition of the real numbers.

[18] Kronecker (1823–1891) was a German mathematician. He worked on number theory, algebra, and logic. Kronecker was hostile to Cantor's theory of sets, since her proofs are often non-constructive in nature, without building concrete examples. Kronecker considered the concept of actual infinity to be absurd.

Set theory is a branch of mathematical logic that studies of objects. Any type of object can be collected into a set.

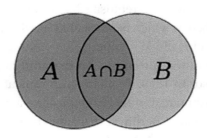

Set theory is applied most often to objects that are relevant to mathematics. The language of set theory can be used to define nearly all mathematical objects. For example, a set of polygons.

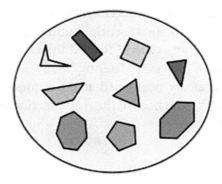

Cantor's set theory came across sharp criticism from a number of well-known contemporary mathematicians. They reminded that before Cantor, all the luminaries of mathematics, from Aristotle, ancient Greek philosopher (380–322 BC) to Gauss (detailed information on Gauss is provided in Section 11.1), considered actual infinity to be an unacceptable scientific concept. The criticism was very aggressive: thus Poincaré called "cantorism" a serious illness that struck mathematical science and expressed the hope that future generations would recover from it. Kronecker attacked Cantor by calling him names: the "scientific charlatan," "apostate," and "corrupter of youth."

Harsh criticism from part of prominent mathematicians was opposed by the worldwide fame and approval of others. In 1904, the Royal Society of London awarded Cantor by the highest mathematical award, the Sylvester Medal[19]. Bertrand Rassell[20] described set theory as "one of the main successes of our era". David Hilbert (detailed information on Hilbert is provided in Section 11.11) called Cantor "a mathematical genius" and declared, "No one can drive us out of the paradise created by Cantor."

Cantor created set theory, which has become a fundamental theory in mathematics. Cantor's work is of great philosophical interest, a fact of which he was well aware. Georg Cantor died on January 6, 1918. He was 72 years old.

---

[19] Sylvester Medal (bronze) awarded for the encouragement of mathematical research and accompanied by a £1,000 prize. It was named in honor of James Joseph Sylvester (1814–1897). He was an English mathematician.

[20] Bertrand Russell (1872–1970) was a British philosopher, mathematician, historian, writer, social critic, political activist, and Nobel laureate in literature in 1950.

11.11 David Hilbert (1862–1943)

"A mathematical theory is not to be considered complete until you have made it so clear that you can explain it to the first man whom you meet on the street."

David Hilbert

David Hilbert (Ref 112) was a German mathematician. He is recognized as one of the most influential and universal mathematicians of the 19th and early 20th centuries.

David was born in Königsberg (Prussia), on January 23, 1862, into the family of Otto Hilbert (father) and Maria Therese Erdtmann (mother). David was six years old when his sister Elsie was born.

In 1872, David entered the Friedrichskolleg Gymnasium and after seven years of study there, he transferred to the more science-oriented Wilhelm Gymnasium. Upon graduation, in 1880, young man enrolled at the University of Königsberg, where he became friends with Hermann Minkowski[21] and Adolf Hurwitz[22]. Together they often made long "mathematical walks", where they actively discussed the

---

[21] Hermann Minkowski (1864–1909) was a German mathematician. He is best known for his work in relativity (1907), in which he showed that his former student Albert Einstein's special theory of relativity (1905) could be understood geometrically as a theory of four-dimensional space-time, since known as the "Minkowski space-time."

[22] Adolf Hurwitz (1859–1919) was a German mathematician who worked on algebra, geometry, analysis, and number theory.

solution of scientific problems. David Hilbert later legalized such walks as an integral part of <u>the education of his students.</u>

In 1885, Hilbert defended his doctoral thesis on the theory of invariants, and the following year became Professor of Mathematics at the University of Königsberg. Hilbert was extremely conscientious about reading lectures and eventually gained a reputation as a brilliant teacher. He remained there from 1886 to 1895.

Felix Klein (1849–1925), a German mathematician and mathematics educator, invited in 1895 David Hilbert to work at the University of Göttingen as the Head of the Department of Mathematics. In this position, he remained 35 years, in fact, until the end of his life. At one time, this department was led by Carl Friedrich Gauss (described in Section 11.1), and Bernhard Riemann (described in Section 11.7).

In 1897, Hilbert published his classic monograph "Report on Numbers" on the theory of algebraic variety[23]. Further, Hilbert sharply changed the themes of his research and in 1899 published "The Foundations of Geometry," which also became <u>classical.</u>

In 1900, at the Second International Mathematical Congress in Paris, Hilbert formulated the famous list of twenty-three unsolved problems, which served as a guideline for the application of the efforts of mathematicians throughout the twentieth century. Then, these problems were not solved. Currently, 16 problems out of 23 have been solved.

In the 1910s, Hilbert created in his modern form a functional analysis by introducing a concept called the *Hilbert space* (named after David Hilbert), which generalizes Euclidean space to an infinite-dimensional case. This theory turned out to be extremely useful in mathematics, and also in many sciences: quantum mechanics, the kinetic theory of gases, and others.

After Hitler came to power in Germany, Hilbert lived in Göttingen aside from university affair. Many of his colleagues, most Jews, were

---

[23] The classical definition of an algebraic variety is the set of solutions of a system of algebraic equations over real or complex numbers. Modern definitions generalize it in various ways but try to preserve the geometric intuition corresponding to this definition.

forced to emigrate. One day, Bernhard Rust, the Nazi Minister of Science, Education, and National Culture, asked Hilbert, "How is mathematics now at the University of Göttingen, after it was freed from Jewish influence?" Hilbert sadly replied, "Mathematics at the University of Göttingen? It is disappeared." David Hilbert died on February 14, 1943 in Göttingen. He was 81 years old.

11.12 Srinivasa Ramanujan (1887–1920)

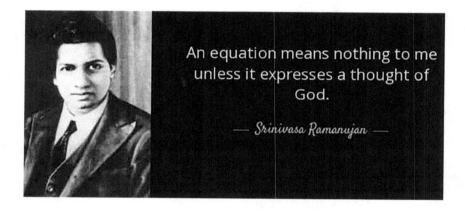

Srinivasa Ramanujan (Ref 113) was an Indian mathematician. He lived during the British Rule in India. Though he had no formal training in pure mathematics, he made substantial contribution to it.

Ramanujan was born on December 22, 1887, at the residence of his maternal grandparents. His father worked as a clerk in a sari shop. His mater was housewife and also sang at a local temple.

In 1889, Ramanujan suffered smallpox, but managed to survive and recover. On October 1, 1892, he was enrolled at the local school. At school, his outstanding abilities in mathematics showed up, and a familiar student from the city of Madras gave him books on trigonometry.

At the age of 14, Ramanujan discovered Euler's sine and cosine formula and was very upset to hear that it was already published.

$$e^{ix} = cosx + isinx$$

At the age of 16, Ramanujan has read and studied closely the book *Synopsis of Pure Mathematics by George Shoobridge Carr*[24]).

Synopsis of Pure Mathematics contains 6165 theorems and formulas, practically without proofs or explanations. Ramanujan had no access to the universities, nor contacts with mathematicians. He was plunged into communication with this book of formulas. Thus, he formed a certain way of thinking, a peculiar style of evidence. During this period, the mathematical fate of Ramanujan was determined.

In 1913, the Cambridge University professor Godfrey Hardy[25] received a letter from Ramanujan. He reported that did not graduate from the university, and after high school was engaged in mathematics independently. The formulas were attached to the letter. Ramanujan asked to publish them if they are interesting, since he himself is poor and does not have sufficient funds to publish.

A lively correspondence began between the Cambridge professor and the Indian clerk. As a result, Godfrey Hardy accumulated about 120 formulas and theorems, some of which were unknown to science. Hardy was impressed by some of Ramanujan's work related to infinite series. One of them is shown below.

$$1 - 5\left(\frac{1}{2}\right)^3 + 9\left(\frac{1 \times 3}{2 \times 4}\right)^3 - 13\left(\frac{1 \times 3 \times 5}{2 \times 4 \times 6}\right)^3 + \cdots = \frac{2}{\pi}$$

The left-hand part of the above equality contains four terms. Let's summate these terms:

The 1st and 2nd terms: $1 - 5 \times (1/2)^3 = 1 - 5(1/8) = 1 - 5/8 = 3/8 = 0.3750$

The 3rd term: $9 \times [(1 \times 3)/(2 \times 4)]^3 = 9 \times [3/8]^3 = 9 \times 0.375^3 = 9 \times 0.052734 = 0.47461$

---

[24] George Shoobridge Carr (1837–1914) was a British mathematician.
[25] Godfrey Hardy (1877–1947) was leading English pure mathematician whose work was mainly in analysis and number theory.

The sum of the 3 terms: $0.3750 + 0.47461 = \underline{0.84961}$

The 4th term:

$-13 \times [(1 \times 3 \times 5)/(2 \times 4 \times 6)]^3 = -13 \times [15/48]^3 = -13 \times 0.3125^3 = -13 \times 0.03052 = \underline{-0.39673}$

The sum of the 4 terms: $0.84961 - 0.39673 = \underline{0.45288}$
The right-hand part of the equality is:

$2/\pi = 0.63662$

The value of the right-hand part of the equality exceeds the sum of the first 4 terms of the left-hand part of the equality by:

$0.63662 - 0.45288 = 0.18374$, or $\underline{28.9\%}$

The sum of the 4 terms (0.45288) in the left-hand part of the equality is less (about 29%) than the value of the right-hand part of the equality. Let's see what happens if we add six more terms.

The 5th term:

$17 \times [(1 \times 3 \times 5 \times 7)/(2 \times 4 \times 6 \times 8)]^3 = 17 \times [105/384]^3 = 17 \times 0.2734375^3 = 17 \times 0.020444 = \underline{0.347548}$

The sum of the 4 terms (0.45288) and the 5th term (0.34748) is:

$0.45288 + 0.34748 = \underline{0.80036}$

In this case, the sum of the 5 terms is greater than the value of the right-hand part of the equality by:

$0.80036 - 0.63662 = 0.16374$, or $\underline{25.7\%}$

As you see, the sum of the 5 terms reduced the difference in comparison of the sum of the 4 terms by: 28.9% − 25.7% = 3.2%

<u>The 6th term:</u> − 21 × [(1 × 3 × 5 × 7 × 9)/(2 × 4 × 6 × 8 ×10)]^3 = − 21 × [945/3840]^3 =

− 21 × 0.014903963 = − 0.312983

<u>The 7th term:</u> 25 × [(1 × 3 × 5 × 7 × 9 × 11)/(2 × 4 × 6 × 8 ×10 × 12)]^3 =

25 × [10395/46080]^3 = 25 × 0.225585938^3 = 25 × 0.011479846 = 0.2870

Summarize the 5 terms (0.80036), the 6th term (− 0.31298), and the 7th term (0.2870):

0.80036 − 0.31298 + 0.2870 = 0.77438

In this case, the sum of the 7 terms is greater than the value of the right-hand part of the equality by:

0.77438 − 0.63662 = 0.13776, or 21.6%

As you see, the sum of the 7 terms reduced the difference in comparison of the sum of the 4 terms by: 28.9% − 21.6% = 7.3%

<u>The 8th term:</u> − 29 × [(1 × 3 × 5 × 7 × 9 × 11 × 13)/(2 × 4 × 6 × 8 ×10 × 12 ×14)]^3 =

− 29 × [135135/645120]^3 = − 29 × 0.209472656^3 = − 29 × 0.009191407 = −0.26655

To the sum of the 7 previous terms (0.77438) add the 8th term (−0.26655) and got:

$$0.77438 - 0.26655 = \underline{0.50783}$$

Now, the sum of the 8 terms is less than the value of the right-hand part of the equality by:

$$0.63662 - 0.50783 = 0.12879, \text{ or } \underline{20.2\%}$$

The sum of the 8 terms reduced the difference in comparison of the sum of the 4 terms by: $28.9\% - 20.2\% = \underline{8.7\%}$

<u>The 9th term</u>: $33 \times [(1\times3\times5\times7\times9\times11\times13\times15) / (2\times4\times6\times8\times10\times12\times14\times16)]^3 =$

$$33 \times [2027025/10321920]^3 = 33 \times 0.196380615^3 =$$
$$33 \times 0.007573486 = \underline{0.24993}$$

To the sum of the 8 previous terms (0.50783) add the 9th term (0.24993) and got:

$$0.50783 + 0.24993 = \underline{0.75776}$$

Now, the sum of the 9 terms (0.75776) is greater than the value of the right-hand part of the equality (0.63662) by:

$$0.75776 - 0.63662 = 0.12114, \text{ or } \underline{19.0\%}$$

The sum of the 9 terms reduced the difference in comparison of the sum of the 4 terms by: $28.9\% - 19.0\% = \underline{9.9\%}$

<u>The 10th term</u>:

$$-37 \times [(1\times3\times5\times7\times9\times11\times13\times15\times17\times19)/(2\times4\times6\times8\times10\times12\times14\times16\times18\times20)]^3 =$$
$$-37 \times [38513475/206438400]^3 = -37 \times 0.186561584^3 =$$
$$-37 \times 0.006493318 = \underline{-0.24025}$$

To the sum of the 9 previous terms (0.75776) add the 10th term (−0.24025) and got:

$$0.75776 - 0.24025 = \underline{0.51751}$$

Now, the sum of the 10 terms (0.51751) is less than the value of the right-hand part of the equality (0.63662) by:

$$0.63662 - 0.51751 = 0.11911, \text{ or } \underline{18.7\%}$$

The sum of the 10 terms reduced the difference in comparison with the sum of the 4 terms by: 28.9% − 18.7% = $\underline{10.2\%}$

Dear readers, if you are interested in continuing this process (adding terms to the left-hand part of the equality and comparing the summation with the right-hand part) you should: each time add the number "4" to the previous coefficient. For example, the coefficient of the 10th term is − 37. The coefficient of the 11th term is + 41, the coefficient of 12th term is − 45, etc.

Hardy and Littlewood[26] began to look at Ramanujan's notebooks. There were many formulas and theorems there.

Hardy saw that some were wrong, others had been discovered, and the rest were new breakthrough. Ramanujan left a deep impression on Hardy and Littlewood. Littlewood commented, "I can believe that he's at least a Jacobi"[27], while Hardy said, <u>"I can compare him only with Euler or Jacobi."</u>

At the insistence of Hardy, Ramanujan at the age of 27, moved to Cambridge. There he was elected a fellow of the Royal Society

---

[26] John Edensor Littlewood (1885–1977) was an English mathematician. He worked on topics relating to analysis, number theory, and differential equations. He and Hardy had a lengthy collaboration.

[27] Carl Gustav Jacob Jacobi (1804–1851) was a German mathematician, who made fundamental contribution to elliptic functions, dynamics, differential equations, and number theory. In 1827, Jacobi became professor at the University of Berlin. He was the first Jewish mathematician to be appointed professor at a German university. In 1836, Jacobi had been elected a foreign member of the Royal Swedish Academy of Sciences.

of London, and at the same time a professor at the University of Cambridge. He was the first Indian to receive such honors. Printed works with his formulas came out one by one, causing surprise, and sometimes bewilderment of colleagues.

In March 1916, Ramanujan was awarded a Bachelor of Science degree by research (this degree was later renamed PhD) for his paper on highly composite numbers, also known as an anti-prime number. It is a positive integer with more divisors than any smaller positive integer has. His paper was more than 50 pages and proved various properties of such numbers. Hardy remarked that this paper was one of the most unusual papers seen in mathematical research at that time, and that Ramanujan showed extraordinary ingenuity in handling it. The concept might have been known to Plato, who set 5040 as it has more divisors than any numbers less than it.

Throughout his life, Ramanujan was plagued by health problems. His health worsened in England during 1914–1918. He was diagnosed with tuberculosis and a severe vitamin deficiency. In 1919, he return to India, and died on April 26, 1920 at the age of 32.

11.13 Andrey Nikolaevich Kolmogorov (1903–1987)

Andrey Nikolaevich Kolmogorov
(1903, Tambov, Russia—1987 Moscow)

- Measure Theory
- Probability
- Analysis
- Intuitionistic Logic
- Cohomology
- Dynamical Systems
- Hydrodynamics
- Kolmogorov complexity

Kolmogorov (Ref 114) was a Soviet mathematician who made significant contribution to the variety of eight mathematical fields that listed above and also: topology, turbulence, classical mechanics, and algorithmic information theory. He was one of the greatest mathematicians of the 20th century.

Kolmogorov was born on April 25, 1903, in Tambov, about 500 km (310 mi) southeast of Moscow. His mother died giving birth to him. He was raised by two of his aunts, who lived near Yaroslavl at the estate of his grandfather, a well-to-do nobleman. Andrey was educated in the village school, and his earliest literary efforts and mathematical papers were printed in the school journal "The Swallow of Spring." Andrey's first mathematical discovery was published in this journal. At the age of five he noticed the regularity in the sum of the series of odd numbers:

$1 = 1 = 1^2$
$1 + 3 = 4 = 2^2$
$1 + 3 + 5 = 9 = 3^2$
$1 + 3 + 5 + 7 = 16 = 4^2$
$1 + 3 + 5 + 7 + 9 = 25 = 5^2$
$1 + 3 + 5 + 7 + 9 + 11 = 36 = 6^2$, etc.

In 1910, Andrey's aunt adopted him, and they moved to Moscow. Andrey was assigned to a private gymnasium Repman, one of the few where boys and girls studied together. Andrey already in those years discovers remarkable mathematical abilities. The teachers did not have time to teach him, and Andrey learned the mathematics himself using the "Brockhaus and Efron Encyclopedic Dictionary."

From the memoirs of Andrey Kolmogorov, "In 1918-1920, life in Moscow was not easy. In schools, only the most persistent were seriously engaged. At that time I had to go for the construction of the Kazan-Yekaterinburg railway. Simultaneously with the work, I continued to study independently, getting ready to take an external program for secondary school. Upon returning to Moscow, I was somewhat disappointed: I was given a certificate of graduation from school, without requiring me to take exams."

In 1920, Andrey Kolmogorov entered the mathematical department of the Moscow University and in parallel the mathematical department of the Mendeleev Chemical and Technological Institute.

In the first months, Kolmogorov passed the exams for the course. And as a second-year student, he gets the right to a "scholarship": "I got the right to 16 kilograms (35.3 lbs.) of bread and 1 kilogram (2.2 lbs.) of butter per month, which, according to that time, meant already complete material well-being."

In 1922, Kolmogorov gained international recognition for constructing a Fourier series[28] that diverges almost everywhere. Around this time, he decided to devote his life to mathematics.

In 1925, Kolmogorov graduated from the Moscow State University. The same year, he published the article "On the principle of the excluded middle," in which he proved that under a certain interpretation, all statements of classical formal logic can be formulated as those of intuitionistic logic.

In 1931, Kolmogorov was awarded the title of Professor of the Moscow State University. In 1933, Kolmogorov published his book "Foundations of the Theory of Probability" laying the modern axiomatic foundations of probability theory and established his reputation as the world's leading expert in this field.

In 1935, Kolmogorov became the first chairman of the Department of probability theory at the Moscow State University. The same year, he earned the degree of Doctor of Physical and Mathematical sciences without defense the dissertation.

From 1935 to 1939, Kolmogorov was a director of the Institute of Mathematics and Mechanics at the Moscow State University.

On January 29, 1939, at the age of 35, Kolmogorov was immediately elected (bypassing the title of corresponding member) as a full member of the USSR Academy of Sciences. He became a member of the Presidium of the Academy.

Shortly before the beginning of the Great Patriotic War (June 22, 1941), Andrey Nikolaevich Kolmogorov was awarded the Stalin Prize for his work on the theory of random processes. During the war, mathematicians of the USSR, on the instructions of the Main Artillery Directorate of the Army, carry out complex work in the field of ballistics and mechanics. Kolmogorov, using his studies on the theory of probability, gives the definition of the most advantageous dispersion of projectiles when firing. After the end of the war, Kolmogorov returned to peaceful research.

By the mid-1960s, the leadership of the USSR Ministry of Education concluded that the system of teaching mathematics in the

---

[28] About this series has been written earlier (see page 198).

Soviet secondary school[29] is in deep crisis and needs reform. It was recognized that only obsolete mathematics is taught in secondary school, and its latest achievements are not covered.

The modernization of the system of mathematical education was carried out by the Ministry of Education of the USSR with the participation of the Academy of Pedagogical Sciences and the Academy of Sciences of the USSR. The management of the Department of Mathematics of the USSR Academy of Sciences recommended for the work on the modernization of Academician A. N. Kolmogorov, who played a leading role in these reforms. Under the leadership of Kolmogorov, programs of teaching were developed, and new textbooks on mathematics for secondary schools were repeatedly published: "Geometry" and "Algebra and the Beginning of Analysis" textbooks for grade 9 and grade 10.

In 1976, A. N. Kolmogorov founded the Department of Mathematical Statistics of the Faculty of Mechanics and Mathematics of Moscow State University.

In 1980, he became the head of the department of mathematical logic and remained in this position until his death. On October 20, 1987, he died in Moscow. Buried at the Novodevichy cemetery. He was 84 years old.

On April 5, 1979, Kolmogorov received a head injury when he entered his entrance, which obviously affected his health. According to the official version, the door with a powerful spring hit Kolmogorov from behind, and the bronze handle hit him in the head. During the last years of his life, Kolmogorov suffered from Parkinson's disease.

---

[29] The term of study at secondary schools was 10 years: primary education (grades 1 to 4), incomplete secondary education (grades 5 to 7), and secondary education (grades 8 to 10). Upon graduation from such school, a graduate obtains secondary education and may apply to take the entry exams for university or institute. If students graduated from secondary school with gold medal, or silver medal, they entered university or institute without entrance examinations. Students who graduated from higher education institutions, obtained the higher education.

Note

My daughter, being a student of the 9th and 10th grades of a secondary school in Kiev, USSR, during 1976–78 academic years, taught mathematics using textbook "Algebra and the Beginning of Analysis." This textbook wrote and edited A. N. Kolmogorov. The book was approved by the USSR Ministry of Education.

If there are teachers of mathematics among the readers of my book, I would recommend them to familiarize themselves with the table of contents of those textbooks.

"Algebra and the Beginning of Analysis," textbook for the 9th grade.

Table of contents:
Chapter I. Principle of Mathematical Induction
Chapter II. Elements of Combinatorics
Chapter III. Real Numbers. Infinite Sequences and their Limits
Chapter IV. Limit of Function and Derivative
Chapter V. Application of Derivative
Chapter VI. Trigonometric Functions, their Charts and Derivatives

"Algebra and the Beginning of Analysis", textbook for the 10th grade.

Table of contents:
Chapter VI. Trigonometric Functions, their Charts and Derivatives (continuation)
Chapter VII. Antiderivative and Integral
Chapter VIII. Exponential, logarithmic and power functions
Chapter IX. Systems of equations and inequalities

I am confident that the United States high school math teachers do not teach topics such as: "Limit of Function and Derivative," "Calculation of Integrals," and "Differential Equations."

11.14 Grigori Yakovlevich Perelman (born on June 13, 1966)

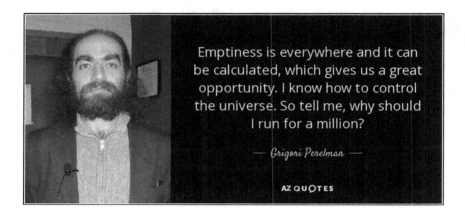

Grigori Yakovlevich Perelman (Ref 115) is a Russian mathematician, known for his contributions to the fields of geometric analysis, Riemannian geometry, and geometric topology.

Grigori was born in Leningrad, Soviet Union (now Saint Petersburg, Russia) on June 13, 1966, to a Jewish family. His father, Yakov Perelman, was an electrical engineer, and in 1993 he immigrated in Israel. Grigori's mother, Lyubov Leybovna Steinholz, remained in St. Petersburg and worked as a teacher of mathematics in vocational school. Lyubov Leybovna plays the violin, and she instilled the love of classical music to Grigori.

Up to grade 9, Grigori studied at a secondary school on the outskirts of Leningrad, and then transferred to the Physics and Mathematics School.

In January 1982, as part of a team of the Soviet schoolchildren, Grigori won the gold medal at the International Mathematical Olympiad in Budapest, receiving a full score for the perfect solution of all problems.

Grigori Perelman graduated from the Physics and Mathematics School but did not receive a gold medal just because he did not pass the physical education exam of the TRP standard—"Ready for labor and defense."

In September 1982, Perelman was enrolled in the Mathematics and Mechanics faculty of the Leningrad State University without exams. There he was particularly influenced by academician Aleksandr Danilovich Aleksandrov[30]. All the years <u>Perelman studied only on "excellent" and received a Lenin scholarship.</u>

In 1987, Perelman graduated with honors from the Leningrad State University and entered as a postgraduate student at the Leningrad branch of the Steklov <u>Mathematical Institute[31]).</u>

One might imagine (Ref 116) that Perelman's achievements would mean that he would be welcomed as a postgraduate student at the Leningrad branch of the Steklov Mathematical Institute with open arms. However, under the academician Ivan Matveevich Vinogradov's[32] leadership the Steklov Mathematical Institute had accepted no Jews and, although it now had a new director, the old policies persisted. Aleksandr Danilovich Aleksandrov appealed to the director[33] of the Leningrad branch of this institute requesting to accept postgraduate student Perelman to work <u>on his dissertation.</u>

---

[30] Aleksandr Danilovich Aleksandrov (1912–1999) was a Soviet/Russian mathematician, physicist, and philosopher. In 1964, he became a full member of the USSR Academy of Sciences, i.e. academician.

[31] Steklov Mathematical Institute is a premier research institute based in Moscow, specialized in mathematics, and a part of the Russian Academy of Sciences. The institute is named after Vladimir Andreevich Steklov (1864–1926). In 1921, he petitioned for the creation of the Institute of Physics and Mathematics. Upon his death the institute was named after him.

[32] Ivan Matveevich Vinogradov (1891–1983) was a Soviet mathematician, one of the creators of modern analytic number theory, and a dominant figure in mathematics in the USSR. Vinogradov enjoyed significant influence in the USSR Academy of Sciences and was regarded as an informal leader of the Soviet mathematicians, not always in a positive way: his anti-Semitic feeling led him to hinder the careers of many prominent Soviet mathematicians of Jewish descent.

[33] The first director of the Leningrad / Saint Petersburg branch of the Steklov Mathematical Institute was a Soviet/Russian scientist, mathematician and theoretical physicist, academician Ludwig Dmitrievich Faddeev (1934–2017). He headed this institute from 1976 to 2000.

Grigori Yakovlevich Perelman has made a significant contribution to Riemannian geometry[34] and geometric topology. In 2003, Perelman proved Thurston's geometrization conjecture[35]). The proof was confirmed in 2006.

In 1990, Perelman defended his thesis *Saddle Surfaces in Euclidean Spaces* for the degree of Candidate of Technical Sciences (equivalent to Ph.D. in USA).

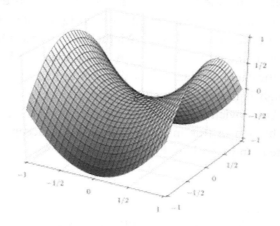

A saddle point (in red) on the graph of $z = x^2 - y^2$

---

[34] About Riemann and his contributions to mathematics are provided in Section 11.7.

[35] Thurston's geometrization conjecture states that certain 3-dimensional topological spaces each have a unique geometric structure that can be associated with them. William Thurston (1946–2012) was an American mathematician. He pioneered in the field of low-dimensional topology. In 1982, he was awarded the Fields Medal[36] for his contributions to the study of 3-manifolds.

[36] The Fields Medal is a prize awarded to two, three, or four mathematicians under 40 years of age at the International Congress of the International Mathematical Union, a meeting that takes place every four years. The Fields Medal is often described as the "Nobel Prize of Mathematics". The name of the award is in honor of Canadian mathematician John Charles Fields (1863–1932).

In November 2002, Perelman posted the first of a series of eprints to the arXiv, in which he claimed to have outlined a proof of the geometrization conjecture, of which the Poincaré conjecture[37]) is a particular case.

The conjecture states:

> "Every simply connected, closed 3-manifold is homeomorphic to the 3-sphere."

In August 2006, Perelman was offered the Fields Medal for his contributions to geometry and his revolutionary insights into the analytical and geometric structure of the Ricci flow[38]), but he declined the award, stating: "I'm not interested in money or fame; I don't want to be on display like an animal in a zoo."

On March 18, 2010, Perelman was awarded a Millennium Prize[39]) for solving the Poincaré conjecture. On June 8, 2010, he did not attend a ceremony in his honor at the Institut Océanographique, Paris to accept his $1 million prize. Perelman refused to accept it. He considered the decision of the Clay Institute unfair for not sharing the prize with Richard S. Hamilton[40]) and stated that "the main reason is my disagreement with the organized mathematical community. I don't like their decisions; I consider them unjust."

Usually, Perelman refused to talk to journalists. Once, in 2012, a journalist managed to reach Perelman. But Perelman told him, "You are disturbing me. I am picking mushrooms."

---

[37] The Poincaré conjecture is the characterization of the 3-sphere, which is the hypersphere that bounds the unit ball in four-dimensional space.

[38] In differential geometry, the Ricci flow is an intrinsic geometric flow. It is a process that deforms the metric of a Riemannian manifold in a way formally analogous to the diffusion of heat, smoothing out irregularities in the metric. The Ricci flow named after Gregorio Ricci-Curbastro (1853–1925). He was an Italian mathematician.

[39] To date, the only Millennium Prize problem to have been solved is the Poincaré conjecture. It was solved by the Russian mathematician Grigori Perelman in 2003.

[40] Richard Streit Hamilton, born 1943. He is professor of Mathematics at Colombia University in the City of New York.

11.15 John Forbes Nash, Jr. (1928–2015)

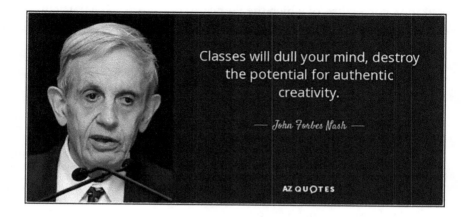

John Nash (Ref 117) was an American mathematician who made fundamental contribution to differential geometry, the study of partial differential equations, and game theory.

John Nash was born on June 13, 1928, in Bluefield, West Virginia. His father, John Forbes Nash, was an electrical engineer for the Appalachian Electric Power Company. His mother, Margaret Virginia, had been a schoolteacher before she was married. Nash attended kindergarten and public school, and he learned from books provided by his parents and grandparents. Nash's parents pursued opportunities to supplement their son's education and arranged for him to take advanced mathematics courses at a local community college during his final year of high school.

John Nash attended Carnegie Institute of Technology (which later became Carnegie Mellon University) through a full benefit of the George Westinghouse Scholarship, initially majoring in chemical engineering. He switched to a chemistry major and eventually to mathematics. After graduating in 1948 (at age 19) with both a Bachelor of Science and a Master of Science in mathematics, Nash accepted a scholarship to Princeton University, where he pursued further graduate studies in mathematics. Nash's adviser and former Carnegie professor Richard Duffin wrote a letter of recommendation for Nash's entrance to Princeton stating, "He is a mathematical genius". Nash was also accepted at Harvard University. The chairman of the mathematics department at Princeton, Solomon Lefschetz (1909–1996), offered him the John S. Kennedy[41] fellowship, convincing Nash that Princeton valued him more. Further, he considered Princeton more favorably because of its proximity to his family in Bluefield.

In September 1948, Nash entered Princeton University where he showed an interest in a broad range of pure mathematics: topology, algebraic geometry, and game theory.

In 1950, Nash earned a Ph.D. degree. The thesis, written under the supervision of doctoral advisor Albert W. Tucker[42], contained the definition and properties of the Nash equilibrium[43], a crucial concept in non-cooperative games[44]. Nash's paper written 45 years later won the Nobel Memorial Prize in Economic Sciences in 1994.

In 1952, Nash published "Real Algebraic Manifolds." It was a groundbreaking work in the area of real algebraic geometry[45].

Nash's work in mathematics includes the Nash embedding theorem, which shows that every abstract Riemannian manifold can be isometrically realized as a submanifold of Euclidean space. He also made significant contributions to the theory of nonlinear parabolic partial

---

[41] John Stewart Kennedy (1830–1909) was an American businessman, financier, and philanthropist.

[42] Albert William Tucker (1905–1995) was a Canadian mathematician who made important contributions in topology, game theory, and non-linear programming.

[43] Nash equilibrium (named after John Nash) is a proposed solution of a non-cooperative game involving two or more players in which each player is assumed to know the equilibrium strategies of the other players, and no player has anything to gain by changing only their own strategy.

[44] A non-cooperative game is a game with competition between individual players, as opposed to cooperative games, and in which alliances can only operate if self-enforcing (e.g. through credible threats). The key distinguishing feature is the absence of external authority to establish rules enforcing cooperative behavior. In the absence of external authority (such as contract law), players cannot group into coalitions and must compete independently.

[45] In mathematics, real algebraic geometry is the study of real algebraic sets, i.e. real-number solutions to algebraic equations with real-number coefficients, and mappings between them (in particular real polynomial mappings).

differential equations and to singularity theory. Mikhail Leonidovich Gromov[46] writes about Nash's work: "Nash was solving classical mathematical problems, difficult problems, something that nobody else was able to do, not even to imagine how to do it. But what Nash discovered in the course of his constructions of isometric embedding is far from 'classical'—it is something that brings about a dramatic alteration of our understanding of the basic logic of analysis and differential geometry. Judging from the classical perspective, what Nash has achieved in his papers is as impossible as the story of his life. His work on isometric immersions opened a new world of mathematics that stretches in front of <u>our eyes in yet unknown directions and still waits to be explored.</u>"

In 1958, Fortune magazine called Nash the rising star of America in the "new mathematics."

In 1978, Nash was awarded the John von Neumann Theory Prize for his discovery of non-cooperative equilibria, now called Nash Equilibria. (The prize named after mathematician John von Neumann is awarded for a body of work, rather than a single piece. The prize was intended to reflect contributions that have stood the test of time. The criteria include significance, innovation, depth, and scientific excellence).

In 1994, Nash received the Nobel Memorial Prize in Economic Sciences, along with John Harsanyi (Hungary, United States) and Reinhard Selten (Germany).

In 1999, Nash won the Leroy P. Steele Prize[47] for his remarkable paper: "The <u>embedding problem for Riemannian manifolds</u>".

On May 19, 2015, a few days before his death, John Nash, along with Louis Nirenberg (a Canadian—American mathematician), were awarded the 2015 Abel Prize by King Harald V of Norway at a ceremony in Oslo.

---

[46] Mikhail Leonidovich Gromov, born December 23, 1943, in Boksitogorsk, town in the Leningrad Region, USSR. He is an American-French-Russian mathematician known for work in geometry, analysis and group theory. He taught in France and in the USA (Professor of Mathematics at New York University).

[47] Leroy P. Steele (1891–1968) is only a piece of the information we found about him.

Mental Illness

In 1958, Nash developed symptoms of schizophrenia. He seemed to believe that all men who wore red ties were part of a communist conspiracy against him. Nash was writing and mailing letters to embassies in Washington, D.C., declaring that they were establishing a new government.

Nash's psychological issues crossed into his professional life when he gave a lecture at Columbia University in 1959. Originally, it was intended to present a proof of the Riemann hypothesis, but the lecture was incomprehensible. Colleagues in the audience immediately realized that something was wrong. In April 1959, he was admitted to McLean Hospital in Belmont, Massachusetts, staying through May of the same year. There, he was diagnosed with paranoid schizophrenia.

In 1961, Nash was admitted to the New Jersey State Hospital at Trenton. Over the next nine years, he spent periods in psychiatric hospitals, where he received both antipsychotic medications and insulin shock therapy. After 1970, he was never committed to a hospital again, and he refused any further medication.

Nash wrote in 1994, "I spent times of the order of five to eight months in hospitals in New Jersey, always on an involuntary basis and always attempting a legal argument for release. And it did happen that when I had been long enough hospitalized that I would finally renounce my delusional hypotheses and revert to thinking of myself as a human of more conventional circumstances and return to mathematical research. In these interludes of, as it were, enforced rationality, I did succeed in doing some respectable mathematical research. But after my return to the dream-like delusional hypotheses in the later 60s I became a person of delusionally influenced thinking but of relatively moderate behavior and thus tended to avoid hospitalization and the direct attention of psychiatrists. Thus further time passed. Then gradually I began to intellectually reject some of the delusionally influenced lines of thinking which had been characteristic of my orientation. This began, most recognizably, with the rejection of politically oriented thinking as essentially a hopeless waste of intellectual effort. So at the present time I seem to be thinking rationally again in the style that is characteristic of scientists."

Personal Life

In 1952, John Nash began a relationship with Eleanor Stier, a nurse he met while admitted as a patient. Nash left Eleanor when she told him of her pregnancy. Soon Nash met a student, Salvadoran beauty Alicia Lardé Lopez-Harrison, a naturalized U.S. citizen from El Salvador. Alicia Lardé graduated from MIT, having majored in physics. They married in February 1957. They had a son John Charles Martin Nash, who earned a Ph.D. degree in Mathematics. Since 1970, Nash continued to work on mathematics and eventually was allowed to teach again.

Death

On May 23, 2015, Nash and his wife were killed in a vehicle accident on the New Jersey Turnpike. They had been on their way home from the airport after a visit to Norway, where Nash had received the Abel Prize. A taxicab driver lost control of the vehicle and struck a guardrail. Both passengers were ejected from the car upon impact. Nash was 86 years old.

John Forbes Nash is the only person to be awarded both the Nobel Memorial Prize in Economic Sciences and the Abel Prize.

APPENDIX 1

INTERNATIONAL STUDENTS PERFORMANCE ON MATHEMATICS IN 2012

Rank	Country	Average Performance	Rank	Country	Average Performance
1	Shanghai, China	613	34	Russia	482
2	Singapore	573	34	Slovak Republic	482
3	Hong Kong, China	561	36	United States	481
4	Taiwan	560	37	Lithuania	479
5	South Korea	554	38	Sweden	478
6	Macau, China	538	39	Hungary	477
7	Japan	536	40	Croatia	471
8	Liechtenstein	535	41	Israel	466
9	Switzerland	531	42	Greece	453
10	Netherlands	523	43	Serbia	449
11	Estonia	521	44	Turkey	448
12	Finland	519	45	Romania	445
13	Canada	518	46	Cyprus	440
13	Poland	518	47	Bulgaria	439
15	Belgium	515	48	United Arab Emirates	434
16	Germany	514	49	Kazakhstan	432

17	Vietnam	511	50	Thailand	427
18	Austria	506	51	Chile	423
19	Australia	504	52	Malaysia	421
20	Ireland	501	53	Mexico	413
20	Slovenia	501	54	Montenegro	410
22	Denmark	500	55	Uruguay	409
22	New Zealand	500	56	Costa Rica	407
24	Czech Republic	499	57	Albania	394
25	France	495	58	Brazil	391
26	United Kingdom	494	59	Argentina	388
27	Iceland	493	59	Tunisia	388
28	Latvia	491	61	Jordan	386
29	Luxembourg	490	62	Colombia	376
30	Norway	489	62	Qatar	376
31	Portugal	487	64	Indonesia	375
32	Italy	485	65	Peru	368
33	Spain	484			

APPENDIX 2

INTERNATIONAL STUDENTS PERFORMANCE ON MATHEMATICS IN 2015

Rank	Country	Average Performance	Rank	Country	Average Performance
1	Singapore	564	36	Lithuania	478
2	Hong Kong, China	548	37	Hungary	477
3	Macau, China	544	38	Slovak Republic	475
4	Chinese Taipei	542	39	Israel	470
5	Japan	532	39	United States	470
6	B-S-J-G, China	531	41	Croatia	464
7	South Korea	524	42	Argentina	456
8	Switzerland	521	43	Greece	454
9	Estonia	520	44	Romania	444
10	Canada	516	45	Bulgaria	441
11	Netherlands	512	46	Cyprus	437
12	Denmark	511	47	United Arab Emirates	427
12	Finland	511	48	Chile	423
14	Slovenia	510	49	Turkey	420
15	Belgium	507	49	Moldova	420

16	Germany	506	51	Uruguay	418
17	Poland	504	51	Montenegro	418
17	Ireland	504	53	Trinidad and Tobago	417
19	Norway	502	54	Thailand	415
20	Austria	497	55	Albania	413
21	New Zealand	495	56	Mexico	408
21	Vietnam	495	57	Georgia	404
23	Russia	494	58	Qatar	402
23	Sweden	494	59	Costa Rica	400
23	Australia	494	60	Lebanon	396
26	France	493	61	Colombia	390
27	United Kingdom	492	62	Peru	387
27	Czech Republic	492	63	Indonesia	386
27	Portugal	492	64	Jordan	380
30	Italy	490	65	Brazil	377
31	Iceland	488	66	Macedonia	371
32	Spain	486	67	Tunisia	367
32	Luxembourg	486	68	Kosovo	362
34	Latvia	482	68	Algeria	362
35	Malta	479	70	Dominican Republic	328

APPENDIX 3

BABYLONIAN NUMERALS

The Babylonians numerals were written in cuneiform script, using a wedge-tipped reed stylus to make an indent on a soft clay tablet, which have been exposed in sunlight to create a permanent document.

The Babylonians were famous for their calculations using a sexagesimal (base-60) positional numeral system.

Only two symbols (𒁹 to count units and 𒌋 to count tens) were used to notate the 59 non-zero digits. These symbols and their values were combined to form a digit quite similar to that of Roman numerals. For example, the combination 𒎙𒁹𒁹𒁹 represented the digit for 23.

The Babylonians invented the abacus (plural **abaci** or **abacuses**), a calculating tool that was used to do addition and subtractions. Originally, abaci were beans or stones moved in grooves in sand or on tablets of wood, stone, or metal. The abacus was in use for centuries before the adoption of the written modern numeral system and was widely used by merchants, traders and clerks in Asia, Africa, and elsewhere. The abaci are often constructed as a bamboo frame with beads sliding on wires.

The sexagesimal system is still in use: time (60 seconds in one minute, 60 minutes in one hour), geometry (360° in a circle, 60° in an angle of an equilateral triangle), trigonometry, and astronomy (minutes and seconds).

APPENDIX 4

VLADIMIR SEMYONOVICH GOLENISHCHEV (1856-1947)

Vladimir Semyonovich Golenishchev was one of the first and most accomplished Russian Egyptologist. He obtained the higher education at the Saint Petersburg University.

In 1884–85 Vladimir Golenishchev organized and financed excavations in Wadi Hammamat (Valley of Many Baths) followed by the research at Tell el-Maskhuta (ruins of ancient city located 62 miles (≈100 km) southwest of Cairo) in 1888–89. During the following two decades he travelled to Egypt more than sixty times and brought back an enormous collection of more than 6,000 ancient Egyptian antiquities, including such priceless artifact as the Moscow Mathematical Papyrus.

In 1909, Golenishchev sold his collection to the Moscow Museum of Fine Art and settled in Egypt. Following the Bolshevik Revolution of 1917, he never returned to Russia, residing in Nice (France) and Cairo (the capital of Egypt).

In Egypt, he established and held the chair in Egyptology at the University of Cairo from 1924 to 1929. Golenishchev died in Nice, France. He was 90 years old. His papers are held at the Pushkin Museum in Paris and also in the Griffith Institute in Oxford, England.

APPENDIX 5

VASILY VASILIEVICH STRUVE (1889-1965)

Vasily Vasilievich Struve was the Soviet authority on the Oriental studies (currently, Middle Eastern studies and East Asian studies), the founder of the Soviet scientific school of research on the ancient Near East history.

In 1907, Struve entered the Department of History at the Faculty of History and Philology of the Petersburg University, where he studied the ancient Greek, Latin, and Egyptian languages. He became proficient in all types of Egyptian hieroglyphic writing. In 1911, Struve graduated from the Petersburg University and continued research work and lecturing at the university. In 1913, he left for Germany for profound studies of the Egyptian language. He returned to Russia in 1916 and became a private docent, then a professor at the Petersburg University in 1920.

From 1918 to 1933, Struve was the head of Department for Art and Culture of Egypt at Hermitage Museum. He studied Akkadian (extinct East Semitic) language, Biblical Hebrew, and Sumerian language. In 1928, Struve earned the Doctor of Historical Sciences degree honoris causa for his scientific achievements.

In 1935, he was elected the full member of the USSR Academy of Sciences, becoming an academician. From 1937 to 1940, he was the head of the Ethnography Institute of the USSR Academy of Sciences. From 1941 to 1950, he was the head of the Institute of Oriental Studies of the USSR Academy of Sciences.

Struve authored around 400 scientific works in his lifetime. He translated the Moscow Mathematical Papyrus and published it in 1930. His scientific research field was not limited to Egyptology. His major scientific works were also on the history of arts of Sumer, Babylonia, Assyria, and other civilizations of the ancient Near East.

Struve died in 1965 and buried in Leningrad (currently, Saint Petersburg, Russia).

APPENDIX 6

ALEXANDER HENRY RHIND (1833-1863)

Alexander Henry Rhind was a Scottish lawyer and Egyptologist. He was born in Wick in the Highlands (town in the far north of Scotland).

Rhind studied at the University of Edinburgh. Suffering from pulmonary disease, he travelled to Egypt (as was the custom amongst wealthy Europeans at the time) where he became fascinated by the ancient culture and antiquities of that country.

He collected material for his book entitled *"Thebes, its Tombs and their Tenants,"* which was published in 1862. A prolific writer with a methodical research style, all through his years in Egypt he continued to battle ill health.

Among the items that he collected was the Rhind Papyrus[1], also known as the Ahmes Papyrus named after its Egyptian scribe. Rhind acquired it in 1858 and following his death it was sold to the British Museum in 1863, along with the similar Egyptian Mathematical Leather Roll. Both are mathematical treatises and both were purchased in the Luxor[2] market, and may have previously been stolen from the Ramesseum[3].

When chemically softened and decoded years afterward, Rhind Papyrus and Mathematical Leather Roll show the Egyptians had computed the value of π as 3.1605, a margin of error of less than one percent.

Rhind has been described as a "young hero," the only "bright shining light of archaeological method and conscience" in the mid-nineteenth century, who plotted the exact location of artefacts and their relationships, the first to do so.

Rhind died in his sleep on July 3, 1863, in Cadenabbia[4] at the age of 30. Along with his 1600-volume library he left a bequest to the Society of Antiquaries of Scotland to fund a lectureship, and the prestigious Rhind Lectures currently hosted by the Society commemorates his name. Rhind directed that a sum from his estate at Sibster[5], Caithness[6], be used for this purpose, once the interests of living parties was extinguished; this eventuated in 1874, 11 years after his death.

---

[1] The Rhind Mathematical Papyrus is one of the known examples of Egyptian mathematicians. It is named after Alexander Henry Rhind a Scottish antiquarian.

[2] Luxor is the Egyptian city that lies atop the ruins of ancient city that the Greeks named "Thebes" and the ancient Egyptians called "Waset".

[3] The Ramesseum is the memorial temple of pharaoh Ramesses II, also known as Ramesses the Great (c.1303–1213 BC).

[4] Cadenabbia is a small community in Lombardy, Italy, in the province of Como, on the west shore of Lake Como.

[5] Sibster is a small farming community.

[6] Caithness is the furthest north county of mainland Scotland.

APPENDIX 7

JAMES GARFIELD AND THE PYTHAGOREAN THEOREM

James Abram Garfield (November 19, 1831–September 19, 1881) was the 20th president of the United States. He was born in Cuyahoga County, Ohio. He spent his youth helping his widowed mother, working at her farm outside of Cleveland, Ohio. James loved outdoors, but he never liked farming. At age sixteen, Garfield ran away to work on the canal boats that shuttled commerce between Cleveland and Pittsburgh. During his six weeks on the boats, he fell overboard fourteen times, caught such a fever that he had to return home. While recovering, Garfield vowed to make his way in the world using brains rather than brawn. While attending Geauga Academy in Chester, Ohio, Garfield supported himself with a part-time teaching position at a district school. From 1851 to 1854, he studied at the Western Reserve Electric Institute in Hiram, Ohio, and earned his living as a school janitor. In 1854, Garfield entered William College in Williamstown, Massachusetts.

In 1856, Garfield graduated from William College and began to teach Greek, Latin, mathematics, history, philosophy, and rhetoric at Western Reserve Electric Institute. In 1867, that Institute was renamed into Hiram College. In addition to teaching, he also practiced law, was a brigadier general in the Civil War, served as Western Reserve's president, and was elected to the U.S. Congress. Garfield was ambidextrous (right-handed on both sides) and liked to entertain his friends by simultaneously writing with one hand in Latin and the other hand in Greek. Garfield contributed an original proof of the Pythagorean Theorem to the hundreds that have been recorded over the centuries. Garfield developed his proof in 1876 while a member of Congress and published it in the April 1, 1876, issue of the *New-England Journal of Education*.

The Pythagorean Theorem has been proved many times, and probably will be proven many more times. But only one proof was made by a United States president. It was James Garfield. He had a lively interest in math and studied even after he decided to run for Congress. One day, he was discussing mathematics with other members of the House of Representatives and came up with a simple and unique proof of the Pythagorean Theorem using a rectangular trapezoid. It was five years before Garfield was elected president.

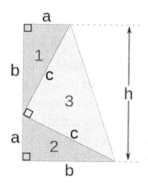

The area (A) of the above trapezoid is calculated by the formula:

$A = (a + b)/2 \times h$
Where $h = a + b$

Replacement of the trapezoid height (h) by (a + b), results in:

$A = (a + b)/2 \times (a + b) = 1/2 (a + b) \times (a + b) =$
$1/2 (a^2 + ab + ab + b^2) = 1/2 (a^2 + 2ab + b^2) =$
$\underline{1/2\, a2 + ab + 1/2\, b2}$

Now we define the area of the trapezoid in terms of the sum of the areas of right triangles 1, 2, and 3:

$A = 1/2\, ab + 1/2\, ab + 1/2\, c^2 = \underline{ab + 1/2\, c2}$

Since the areas of right triangles and the area of trapezoid are equal, we can write:

$1/2\, a^2 + \underline{ab} + 1/2\, b^2 = \underline{ab} + 1/2\, c^2$

Because we have *ab* on either side, they cancel each other out:

$1/2\, a^2 + 1/2\, b^2 = 1/2\, c^2$

By multiplying it all buy 2, we have: $a^2 + b^2 = c^2$

Undoubtedly, many people remember the Pythagorean Theorem as the old problem, wherein the sum of the squares of the legs ($a^2 + b^2$) is equal to the square of the hypotenuse c^2 of a right triangle.

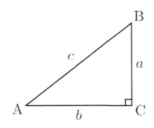

On July 2, 1881, James Garfield was shot twice by Charles Guiteau: one bullet glanced off his arm while the other pierced his back, shattering a rib and embedding itself in his abdomen. It happened at the Baltimore and Potomac Railroad Station in Washington, D.C.

President Garfield was preparing to depart for Williams College where he was scheduled to deliver a speech. Seriously wounded, Garfield was taken upstairs to a private office at the station and then at the White House. The physician who took charge at the station and then at the White House was an old friend of Garfield, Doctor Willard Bliss, a noted physician and surgeon. He and a dozen doctors were soon probing the wound with unsterilized fingers and instruments. At his request, Garfield was taken back to the White House, where he remained until September 6, and then he was moved to a New Jersey seaside retreat to recuperate. James Garfield died in Elberon, New Jersey on September 19, 1881, of infection and internal hemorrhage. According to some historians and medical experts, Garfield might have survived his wounds if the doctors attending him had at their disposal today's medical research, techniques, and equipment.

James Garfield was the second of four presidents to be assassinated, following Abraham Lincoln and preceding William McKinley and John F. Kennedy. His Vice President, Chester A. Arthur, succeeded Garfield as President.

Charles Julius Guiteau (1841–1882) was an American writer and lawyer who was convicted of the assassination of James Garfield. Guiteau was offended by Garfield's rejections of his various job applications. Guiteau was sentenced to death for the crime and was hanged on June 30, 1882.

APPENDIX 8

FIBONACCI NUMBERS IN PYTHAGOREAN TRIPLES

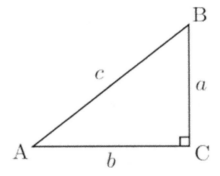

Euclid (described in Section 4.5) proved that the Pythagorean triples are infinite and can be generated by three formulas. In modern mathematical notation, following two formulas (1) and (2) are used to calculate leg BC (a) and leg AC (b) of a right triangle ABC:

$a = m^2 - n^2$ (1)
$b = 2m \times n$ (2)

Hypotenuse AB (c) of a right triangle ABC is calculated by formula (3):

$c = m^2 + n^2$ (3)

The integer m is greater than the integer n ($m > n > 0$). The minimum value of m is 2 and the minimum value of n is 1.

The Fibonacci numbers are the numbers in the following integer sequence, called the Fibonacci sequence, and characterized by the fact that every number after the first two is the sum of the two preceding ones:

F_0	F_1	F_2	F_3	F_4	F_5	F_6	F_7	F_8	F_9	F_{10}	F_{11}	F_{12}	F_{13}	F_{14}	F_{15}	F_{16}	F_{17}
0	1	1	2	3	5	8	13	21	34	55	89	144	233	377	610	987	1597

F_{18}	F_{19}	F_{20}	F_{21}	F_{22}	F_{23}	F_{24}	F_{25}	F_{26}	F_{27}	F_{28}
2584	4181	6765	10946	17711	28657	46368	75025	121393	196418	317811

F_{29}	F_{30}	F_{31}	F_{32}	F_{33}	F_{34}	F_{35}	F_{36}
514229	832040	1346269	2178309	3524578	5702887	9227465	14930352

Conjecture 1:

Let m and n are the Fibonacci numbers entered as one after the other: F_3-F_2; F_4-F_3; F_5-F_4; F_6-F_5; etc. Calculate a, b, and c for the corresponding right triangle using equations (1), (2) and (3). Then c, the hypotenuse of the corresponding right triangle, will always be a Fibonacci number. See Table 1 for examples. Numbers in red are the Fibonacci numbers.

Table 1

Integers		Legs		Hypotenuse
m	n	a	b	c
2	1	3	4	5
3	2	5	12	13
5	3	16	30	34
8	5	39	80	89
13	8	105	208	233
21	13	272	546	610
34	21	715	1428	1597
55	34	1869	3740	4181

89	55	4896	9790	10946
144	89	12815	25632	28657
233	144	33553	67104	75025
377	233	87840	175682	196418
610	377	229971	459940	514229
987	610	602069	1204140	1346269
1597	987	1576240	3152478	3524578

Note: there are only two Fibonacci numbers 3 and 5 for leg *a*.

Conjecture 2:

Let integer *m* and integer *n* are the Fibonacci numbers entered as one through the other: F4-F2; F5-F3; F6-F4; F7-F5; etc. Calculate *a*, *b*, and *c* for the corresponding right triangle using equations (1), (2) and (3). Then *a*, the longer leg of the corresponding right triangle, will always be a Fibonacci number. See Table 2 for examples.

Table 2

Integers		Legs		Hypotenuse
m	*n*	*a*	*b*	*c*
3	1	8	6	10
5	2	21	20	29
8	3	55	48	73
13	5	144	130	194
21	8	377	336	505
34	13	987	884	1325
55	21	2584	2310	3466
89	34	6765	6052	9077
144	55	17711	15840	23761
233	89	46368	41474	62210
377	144	121393	108576	162865
610	233	317811	284260	426389

| 987 | 377 | 832040 | 744198 | 1116298 |

Conjecture 3:

Let integer m and integer n are the Fibonacci numbers entered as one through the two: F5-F2; F6-F3; F7-F4; F8-F5; etc. Calculate a, b, and c for the corresponding right triangle using equations (1), (2) and (3). In this case, the numbers of the legs and the hypotenuse are not the Fibonacci numbers. See Table 3 for examples.

Table 3

Integers		Legs		Hypotenuse
m	n	a	b	c
5	1	24	10	26
8	2	60	32	68
13	3	160	78	178
21	5	416	210	466
34	8	1092	544	1220
55	13	2856	1430	3194
89	21	7480	3738	8362
144	34	19580	9792	21892
233	55	51264	25630	57314
377	89	134208	67106	150050
610	144	351364	175680	392836
987	233	919880	459942	1028458
1597	377	2408280	1204138	2692538
2584	610	6304956	3152480	7049156

APPENDIX 9

SIR ISAAC NEWTON

In the International Metric System (Ref 85, pp.76–82), the unit of force, the **newton**, is named after Isaac Newton, outstanding English physicist, mathematician, astronomer, and philosopher.

Newton was born on December 25, 1642 (by the Julian calendar) or on January 4, 1643 (by the Gregorian calendar) in the village of Woolsthorpe, about 200 kilometers north of London.

Newton's father (also named Isaac Newton) had died three months before his son was born. When Isaac was two years old, his mother Hanna Ayscough remarried Barnabas Smith, the minister of the church at North Witham, a nearby village. Isaac was left in the care of his grandmother. Upon the death of his stepfather in 1653, Isaac lived in an extended family: his mother, grandmother, one half-brother, and two half-sisters.

Isaac Newton began his schooling in the village school in Woolsthorpe. At the age of twelve, he was sent to Grantham Grammar School. In this school Isaac has shown little promise in academic work. The school reports described him as "idle" and "inattentive."

Very interesting facts about Isaac Newton's school years are described in the book *The Founders of Calculus* (in Russian), by the Soviet mathematician Leon Freiman: "At the beginning Newton was, in fact, a mediocre student. He was helped to become a great student by the lucky incident. One of his peers was bullying and beating him up. The direct revenge was excluded, since the enemy was bigger and stronger. So, Isaac decided to surpass his opponent and began to study more diligently, becoming the best student in his school. One of the mathematicians said that we should be very thankful to the boy who beat student Isaac for the genius of great mathematician Newton. Nobody has acted better with his fists." (Translated from Russian by E. Isakov).

By that time Isaac's mother had reasonable wealth and property, and she wanted her eldest son to manage her estate. Isaac was taken away from school, but soon showed that he had no talent, or interests in managing the estate.

Newton's uncle Willian Ayscough decided that Isaac should prepare for entering university and persuaded his sister Hanna that this was the right thing to do. Isaac was allowed to return to the Grantham Grammar School in 1660 to complete his school education.

At Grantham Isaac lodged with the local apothecary William Clarke and eventually became engaged to Clarke's stepdaughter Anne Storey, before he left for Cambridge University at age of 19.

As Newton became preoccupied in his studies, the romance cooled, and Anne Storey married somebody else. It is said he kept a warm memory of this love, but Newton had no other recorded "sweethearts" and never married.

In June 1661, Newton was admitted to Trinity College, Cambridge, where his uncle Willian Ayscough had studied. At the time, the college's teachings were based on the philosophy of Aristotle (384–322 BC), but Newton preferred to study the more advanced ideas of modern philosophers such as René Descartes (1596–1650), Thomas Hobbes (1588–1679), and Robert Hooke (1635–1703). Newton studied Copernican astronomy of Galileo Galilei (1564–1642) and optics of Johannes Kepler (1571–1630).

Newton's deep mathematical studies began when Isaac Barrow (1630–1677), an English mathematician, was selected as the first occupier of the Lucasian chair of mathematics at the University of Cambridge in 1663. Several years later Barrow admitted Isaac Newton's mathematical genius among his students.

In April 1664, Newton was elected a scholar and in April 1665, received his bachelor's degree. In the summer of 1665, the University of Cambridge was closed down as a precaution against the Great Plague. At that time Newton's scientific genius had still not emerged but did so suddenly when he had to return to Grantham. There, Newton began extensively advances in mathematics, optics, physics, and astronomy. In a period of less than two years (1665–1667) he laid the foundation for differential and integral calculus, described his "method of fluxions" and "inverse method of fluxions" in 1671, but published the manuscript only in 1693.

Newton and Leibniz developed the theory of calculus independently using different notations, but a bitter dispute between them sprang up publicly, privately, and had lasted for many years. It didn't stop even after Leibniz's death in 1716.

Newton returned to the University of Cambridge as soon as it was reopened in 1667 after the plague. In October he was elected to a minor fellowship at Trinity College and after receiving his master's degree, he was elected to a major fellowship in July 1668.

In 1669, Isaac Barrow resigned the Lucasian chair to devote himself to divinity. He acknowledged superior abilities of Isaac Newton and recommended him to be appointed in his place.

Newton's first work as Lucasian Professor was on optics that he lectured from 1670 to 1672. During this period he investigated the refraction of light and proved that a beam of sunlight passing through a glass prism decomposes into a spectrum of colors, and that a second prism recomposes multicolored spectrum into white light.

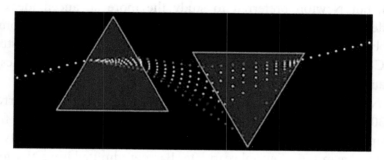

Based on his investigation, Newton concluded that any refracting telescope would suffer from the dispersion of light into colors and constructed a reflecting telescope to bypass that problem. Such telescope had a six-inch-diameter mirror, which Newton ground and polished by himself. In 1671, Newton demonstrated his telescope to Royal Society.

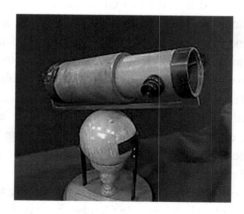

In 1672, Newton was elected a fellow of the Royal Society, and in the same year he published his first scientific paper on light and color. The paper was well received by many scientists, except Robert Hooke and Christiaan Huygens. They objected to Newton's attempt to prove that light consists of the motion of small particles (corpuscles) rather than waves. Newton's corpuscular theory of light reigned until the wave theory was revived in the 19th century. Today's quantum mechanics recognizes a "wave-particle duality." Newton's greatest achievement was his work in physics and celestial mechanics, which culminated in the theory of universal gravitation. Edmond Halley (1656–1742), an English astronomer, mathematician, and physicist, persuaded Newton to write about his new physics and its application to astronomy. In 1687, Newton published his *Mathematical Principals of Natural Philosophy* (commonly known as the *Principia*). In this book, Newton presented his three universal laws of motion.

The **first law** states that every object will remain at rest or in uniform motion in a straight line unless compelled to change its state by the action of an external force. This is normally taken as the definition of **inertia**.

The **second law** explains how the velocity of an object changes when it is subjected to an external force. The law defines a **force** to be equal to change in **momentum** (mass times velocity) per change in time. For an object with a constant mas, this law states that the **force** (F) is the product of an object's **mass** (m) and its **acceleration** (a):

$$F = m \times a$$

The **third law** states that for every action (force) in nature, there is an equal and opposite reaction. In other words, if object **A** exerts a force on object **B**, then object **B** also exerts an equal force on object **A**.

The *Principia* is recognized as the greatest scientific book ever written. Newton analyzed the motion of bodies in resisting and non-resisting media under the action of centripetal forces. The results were applied to orbiting bodies, projectiles, pendulums, and free-fall near the Earth.

Newton used the Latin word *gravitas* (weight) for the force that would become known as **gravity** and defined the law of universal gravitation: *All matter attracts all other matter with the force proportional to the product of their masses and inversely proportional to the square of the distance between them.* This law is expressed by the following equation:

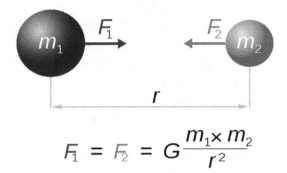

$$F_1 = F_2 = G\frac{m_1 \times m_2}{r^2}$$

Where

F1 and F2 are the forces between the masses
G is the gravitational constant
m1 is the first mass (kg)
m2 is the second mass (kg)
r is the distance between the centers of the masses

Newton explained a wide range of previously unrelated phenomena: the eccentric orbits of comets, the tides and their variations, the precession of the Earth's axis, and motion of the Moon as disturbed by the gravity of the sun.

In January 1689, the University of Cambridge elected Newton as one of their two members to the Convention Parliament for his opposition against King Lames II (ruled from 1685 to 1688) attempt to make universities as Catholic institutions. Newton was a Member of Parliament till 1690 and sat again in 1701–1702. His only recorded comments, while in Parliament, were to complain about a cold draft in the chamber and request that the window be closed.

In 1696, Newton moved to London to take up a government position of Warden of the Royal Mint and became Master of the Mint in 1699, an office he retained to his death. For many people a position such as Master of the Mint would have been treated as a reward for their scientific achievements, but Newton took it seriously, exercising his power to reform the currency and to prevent counterfeiting of the coinage. In 1703, Newton became president of the Royal Society, being annually reelected for the rest of his life. He was knighted in 1705 by Queen Anne, the first scientist to be so honored for his work. Newton died on March 31, 1727, at age 84. He was the first scientist given the honor of burial in Westminster Abbey.

Newton's achievements in physics and mathematics made him an international leader in scientific research. His calculus became vitally important to the development of further scientific theories. He unified many of the physical facts that have been discovered earlier into a satisfying system of laws. For this reason, Isaac Newton is considered one of the greatest scientists, ranking number 1 or 2 with Albert Einstein.

REFERENCES

1. https://nces.ed.gov/surveys/pisa/educators.asp
2. https://en.wikipidea.org/wiki/History_of_ mathematics
3. https://en.wikipidea.org/wiki/Babylonial_numerals
4. https://en.wikipidea.org/wiki/Moscow_Mathematical_Papyrus
5. https://en.wikipidea.org/wiki/Berlin_Papyrus_6619
6. https://en.wikipidea.org/wiki/Rhind_Mathematical_Papyrus
7. https://en.wikipidea.org/wiki/Thales
8. https://en.wikipidea.org/wiki/Thales%27_theorem
9. https://en.wikipidea.org/wiki/Pythagoros
10. Head, Angie. *Pythagorean Theorem.* The University of Georgia, Department of Mathematics Education
11. Abad, Paul and Jack. *The Hundred Greatest Theorems.* http://pirate.shu.edu/~kahlnath/Top100.html
12. http://www.storyofmathematics.com/greek_plato.html
13. https://en.wikipidea.org/wiki/Eudoxus_of_Cnidus
14. https://en.wikipidea.org/wiki/Euclid
15. https://en.wikipidea.org/wiki/Euclid%27s_Elements
16. https://en.wikipidea.org/wiki/Euclidean_geometry
17. https://en.wikipidea.org/wiki/Archimedes
18. https://keisan.casio.com/exec/system/1223282135
19. https://en.wikipidea.org/wiki/Apollonius_of_Perga
20. https://en.wikipidea.org/wiki/Hipparchus
21. https://en.wikipidea.org/wiki/Hero_of_Alexandria
22. https://en.wikipidea.org/wiki/Ptolemy%27s_theorem
23. https://en.wikipidea.org/wiki/Diophantus
24. https://en.wikipidea.org/wiki/Hypatia

25. https://en.wikipidea.org/wiki/Liu_Xin
26. https://en.wikipidea.org/wiki/Zhang_Heng
27. https://en.wikipidea.org/wiki/Liu_Hui
28. https://en.wikipidea.org/wiki/Liu_Hui%27s_%CF%80_algorithm
29. https://en.wikipidea.org/wiki/Zu_Chongzhi
30. kaleidoscope.cultural-china.com/en/135Kaleidoscope5316.html
31. https://en.wikipidea.org/wiki/Yang_Hui
32. https://en.wikipidea.org/wiki/Magic_square
33. https://en.wikipidea.org/wiki/Magic_circle
34. https://en.wikipidea.org/wiki/Indian_mathematics
35. https://en.wikipidea.org/wiki/Aryabhata
36. https://en.wikipidea.org/wiki/Var%C4%81hamihira
37. https://en.wikipidea.org/wiki/Brahmagupta
38. https://en.wikipidea.org/wiki/Brahmagupta_theorem
39. https://en.wikipidea.org/wiki/Bh%C4%81skara_I
40. https://en.wikipidea.org/wiki/Mah%C4%81v%C4%ABra_ (mathematician)
41. http://www-groups.dcs.st-and.ac.uk/history/Biographies/Mahavira.html
42. https://en.wikipidea.org/wiki/Bh%C4%81skara_II
43. http://www-groups.dcs.st-and.ac.uk/history/Biographies/Bhaskara_II.html
44. https://en.wikipidea.org/wiki/Madhava_of_Sangamagrama
45. https://en.wikipidea.org/wiki/Nilakantha_Somayaji
46. http://www-history.mcs.st-andrews.ac.uk/Biographies/Nilakantha.html
47. https://en.wikipidea.org/wiki/Jey%E1%B9%A3%E1%B9%ADhadeva
48. https://en.wikipidea.org/wiki/Muhammad_ibn_Musa_al-Khwarizmi
49. https://en.wikipidea.org/wiki/Ab%C5%AB_K%C4%81mil_Shuj%C4%81%CA%BF_ibn_...
50. http://www-groups.dcs.st-and.ac.uk/history/Biographies/Abu_Kamil.html
51. https://en.wikipidea.org/wiki/Al-Karaji

52. http://www-history.mcs.st-andrews.ac.uk/Biographies/Al-Karaji.html
53. https://en.wikipidea.org/wiki/Alhazen
54. https://en.wikipidea.org/wiki/Omar_Khayyam
55. https://en.wikipidea.org/wiki/Al-Samawal_al-Maghribi
56. http://www-history.mcs.st-andrews.ac.uk/Biographies/Al-Samawal.html
57. https://en.wikipidea.org/wiki/Nasir_al-Din_al-Tusi
58. https://en.wikipidea.org/wiki/Jamsh%C4%ABd_al-K%C4%81sh%C4%AB
59. https://en.wikipidea.org/wiki/Fibonacci
60. https://en.wikipidea.org/wiki/Golden_ratio
61. https://en.wikipidea.org/wiki/Golden_spiral
62. https://en.wikipidea.org/wiki/Thomas_Bradwardine
63. https://en.wikipidea.org/wiki/William_of_Heytesbury
64. https://en.wikipidea.org/wiki/Nicole_Oresme
65. https://en.wikipidea.org/wiki/Piero_della_Franceska
66. https://en.wikipidea.org/wiki/Luka_Pacioli
67. https://en.wikipidea.org/wiki/Sipione_del_Ferro
68. https://ru.wikipidea.org/wiki/%D0%94%D0%B5%D/D
69. http://www-history.mcs.st-andrews.ac.uk/Biographies/Ferro
70. https://en.wikipidea.org/wiki/Niccol%C3%B2_Font
71. BIOGRAPHY: Niccolò Fontana Tartaglia
72. https://en.wikipidea.org/wiki/Gerolamo_Cardano
73. https://en.wikipidea.org/wiki/Rafael_Bombelli
74. https://en.wikipidea.org/wiki/imaginary_number
75. https://en.wikipidea.org/wiki/Fran%C3%A7ois_Vi%
76. http://www-groups.dcs.st-and.ac.uk/history/Biographies/Pitiscus
77. https://en.wikipidea.org/wiki/Bartholomaeus_Pitiscus
78. https://en.wikipidea.org/wiki/Scientific_revolution
79. https://en.wikipidea.org/wiki/John_Napier
80. https://en.wikipidea.org/wiki/Ren%C3%A9_Descart
81. https://en.wikipidea.org/wiki/Pierre_de_Fermat

82. Nikiforovsky, Viktor Arsenievich and Freiman, Leon Semënovich. *The Birth of New Mathematics* (in Russian), Publisher "Nauka" ("Science"), Moscow, 1976
83. https://en.wikipidea.org/wiki/Zerah_Colburn (mental calculator)
84. https://en.wikipidea.org/wiki/Blaise_Pascal
85. Isakov, Edmund. *International System of Units (SI)*. Industrial Press, Inc. 32 Haviland Street, South Norwalk, Connecticut, 06854
86. https://en.wikipidea.org/wiki/Gottfried_Wilhelm_Leibniz
87. www.storyofmathematics.com/17th_leibniz.html
88. https://en.wikipidea.org/wiki/Isaac_Newton
89. www.storyofmathematics.com/17th_newton.html
90. https://en.wikipidea.org/wiki/Leonhard_Euler
91. W Эйлер, Леонард – Википедия
92. http://www-history.mcs.st-andrews.ac.uk/Biographies/Lagrange.html
93. https://en.wikipidea.org/wiki/Joseph-Louis_Lagrange
94. https://en.wikipidea.org/wiki/Pierre-Simon_Laplace
95. https://www.storyofmathematics.com/19th.html
96. https://www.storyofmathematics.com/20th.html
97. https://en.wikipidea.org/wiki/Carl_Friedrich_Gauss
98. https://www.khanacademy.org.../what-is-modular-arithmetic
99. https://en.wikipidea.org/wiki/Modular_arithmetic
100. http://www.storyofmathematics.com/19th_gauss.html
101. https://en.wikipidea.org/wiki/Augustin-Louis_Cauchy
102. https://en.wikipidea.org/wiki/Nikolai_Lobachevsky
103. https://en.wikipidea.org/wiki/Niels_Henrik_Abel
104. https://en.wikipidea.org/wiki/Évariste_Galois
105. https://www.storyofmathematics.com/19th galois.html
106. https://en.wikipidea.org/wiki/János_Bolyai
107. https://en.wikipidea.org/wiki/Bernhard Riemann
108. https://en.wikipidea.org/wiki/Sofia_Kovalevskaya
109. https://www.agnesscott.edu/lriddle/women/kova.htm
110. https://en.wikipidea.org/wiki/Henri_Poincar%C3A9
111. https://en.wikipidea.org/wiki/Georg_Cantor

112. https://en.wikipidea.org/wiki/David_Hilbert
113. https://en.wikipidea.org/wiki/Srinivasa_Ramanujan
114. https://en.wikipidea.org/wiki/Andrey_Kolmogorov
115. https://en.wikipidea.org/wiki/Grigori_Perelman
116. http://www.-history-mcs.st-andrews.uk/Biographies/Perelman.html
117. https://en.wikipidea.org/wiki/John_Forbes_Nash_Jr.#Awards

AFTERWORD

Sir Francis Bacon (1561–1626), English philosopher, scientist, and author said:

> "Knowledge gives you power
> Power gives you control,
> Control gives you freedom,
> Freedom gives you potential,
> Potential gives you endless possibilities."

ABOUT THE AUTHOR

The author is not a mathematician, and when his friends ask why he wrote this book, he says, "I did it because I love mathematics."

Edmund Isakov was born in the small town of Nezhin, (Ukraine, USSR), on December 31, 1932, into a Jewish family. Having "Jewish" as a nationality allowed the Soviet government secretly and unofficially implement the "soft forms" of discrimination and restriction in civil rights.

For the Soviet Jews it was next to impossible to be admitted to colleges and universities, get a job, promotions, receive the state awards, honorary titles, trip abroad, and much more.

Edmund Isakov graduated from the Kiev high school # 34 in June 1951. He was nominated for a Gold medal but received a Silver medal. It was just only one of many occasions of anti-Semitism in the Soviet Union.

The gold and silver medals gave the right to be admitted to the institutes and universities without entrance examinations.

In June 1956, Isakov graduated from the Kiev Polytechnic Institute with a master's degree in a specialty "Engineer of the mining electro-mechanical machines."

From 1957 to 1975, he worked at the Institute for Superhard Materials (ISM) in Kiev, USSR.

From 1961 to 1966, Edmund Isakov developed a series of carbide tools for drilling holes in building structures.

Since 1965, these tools have been manufactured at the Kamenets-Podolsk plant of the Ukrainian SSR for all construction companies in the Soviet Union.

In 1971, carbide tools for making holes in building materials was approved by the State Committee of Standards of the Council of Ministers of the USSR.

In 1972, Edmund Isakov successfully defended his dissertation for the degree of Candidate of Technical Sciences, which is equivalent to Doctor of Philosophy, Ph.D. The author, in his dissertation, presented the results of research on the drilling of reinforced concrete by carbide tools that he developed. These tools allowed to drill reinforced concrete much faster than those primitive tools made by local workshops.

For 18 years working at the ISM, Dr. Isakov was granted ten USSR Inventors' Certificates (equivalent to patents) for carbide and diamond tools for drilling in the building materials, including reinforced concrete. During all that time, Dr. Isakov has never been on business trips abroad, because Director of ISM was forbidden to send Jewish scientists abroad.

Life in the Soviet Union became unbearable for Jewish families. Anti-Semitism has manifested itself in many areas, from domestic relations to government policy.

In June 1978, the Isakovs' daughter graduated from the high school with "excellent" marks in all subjects, and was nominated for the gold medal, but she was deprived of such.

Higher education in the Soviet Union was free of charge. It was impossible to be admitted to the Kiev University without high-level acquaintance or bribes. However, there was a belief in their daughter's excellent knowledge and justice. But the miracle did not happen. She was not accepted.

In August 1978, the Isakov family decided to leave the Soviet Union and began to collect the necessary documents to leave this anti-Semitic country forever.

On October 16, 1979, the Isakovs emigrated from the Soviet Union with the hope of being accepted by the government of the United States and becoming citizens of that great country.

Edmund, Yelena, and Marina Isakov landed at the John F. Kennedy International Airport on January 24, 1980.

In July 1980, Dr. Isakov began working as a lab technician in a small company receiving a salary of $5 per hour. Since May 1981, Isakov was sending out resumes to companies dealing with carbide and diamond cutting tools. The first company that responded to his resume was Kennametal Inc., and it invited Edmund Isakov for an interview on June 20, 1981. The company offered him a research engineer position with the initial salary of $27,000 a year. Edmund Isakov accepted the offer of the company and started work on July 20, 1981.

In 1985, Dr. Isakov was granted his first US patent "Two-prong rotary bit, especially for use with roof drilling, and insert therefor". Six more US patents he awarded before retirement. He retired from Kennametal on November 30, 1999.

Dr. Edmund Isakov has had a long and distinguished career in metalcutting. He is especially known for his work in the research, development, and applications of cutting tools for milling, turning, drilling, and boring. He has authored numerous articles on metalcutting in the magazine "Cutting Tool Engineering", and the interactive calculations software Advanced Metalcutting Calculators in both U.S. and metric units (published by Industrial Press). Two articles *Disc Springs* and *Statistical Analysis of Manufacturing* were included in the "Machinery's Handbook", 27th to 30th Editions (published by Industrial Press). The article *Boring* was included in the "Machinery's Handbook, 31st Edition (published by Industrial Press).

Isakov's publications also include four books: *Mechanical Properties of Work Materials* (published by Modern Machine Shop); *Engineering Formulas for Metalcutting, Cutting Data for Turning of Steel*, and *International System of Units SI* (published by Industrial Press).

Lorelle Young, former President U.S. Metric Association, Inc. highly appreciated this book: "*The International System of Units (SI) is an extraordinary book. This volume stands out among other books I have read on the metric system. Not only is it very readable but it is impressive in the scope and depth of coverage of the history, present status of the metric system in the United States and in the world, and outstanding individuals responsible for the science that created metric units…Anyone who reads this book will grow in appreciation of this remarkable measurement system.*"

Dear readers, if you found this book interesting and it provoked your desire to learn mathematics, then the author fulfilled his duty.

The manuscript "Mathematics And Its Creators" was completed in August 2021.

The author can be contacted at edmundisakov2017@gmail.com

Printed in the USA
CPSIA information can be obtained
at www.ICGtesting.com
LVHW021319260923
759343LV00004B/136